Basics of Metal Mining Influenced Water

Basics of Metal Mining Influenced Water

Editor

Harish Kumar

Basics of Metal Mining Influenced Water

Edited by **Harish Kumar**

Printed in 2017

ISBN: 978-1-68117-473-0

Library of Congress Control Number: 2015936591

© 2016 by
SCITUS Academics LLC,
616, Corporate Way, Suite 2, 4766,
Valley Cottage, NY 10989

www.scitusacademics.com

Contents

Preface

Basics of Metal Mining Influenced Water is a must-read for planners, regulators, consultants, land managers, students, researchers, or others concerned about the environmentally sound management of metal mine wastes and drainage quality. Planning a new mine in today's increasingly contentious regulatory and political environment demands a different philosophy. Basics of Metal Mining Influenced Water takes an innovative, holistic approach by considering all aspects of the mine life cycle, including closure. Written by a team of experts from state and federal governments, academia, and the mining industry, Basics of Metal Mining Influenced Water also discusses the major physical and chemical relationships between mining, climate, environment, and mine waste drainage quality.

Editor

Heavy Metals in Contaminated Soils: A Review of Sources, Chemistry, Risks and Best Available Strategies for Remediation

Raymond A. Wuana[1] and Felix E. Okieimen[2]

[1]Analytical Environmental Chemistry Research Group, Department of Chemistry, Benue State University, Makurdi 970001, Nigeria

[2]Research Laboratory, GeoEnvironmental & Climate Change Adaptation Research Centre, University of Benin, Benin City 300283, Nigeria

ABSTRACT

Scattered literature is harnessed to critically review the possible sources, chemistry, potential biohazards and best available remedial strategies for a number of heavy metals (lead, chromium, arsenic, zinc, cadmium, copper, mercury and nickel) commonly found in contaminated soils. The principles, advantages and disadvantages of immobilization, soil washing and phytoremediation techniques which are frequently listed among the best demonstrated available technologies for cleaning up heavy metal contaminated sites are presented. Remediation of heavy metal contaminated soils is necessary to reduce the associated risks, make the land resource available for agricultural production, enhance food security and scale down land tenure problems arising from changes in the land use pattern.

INTRODUCTION

Soils may become contaminated by the accumulation of heavy metals and metalloids through emissions from the rapidly expanding industrial areas, mine tailings, disposal of high metal wastes, leaded gasoline and paints, land application of fertilizers, animal manures, sewage sludge, pesticides, wastewater irrigation, coal combustion residues, spillage of petrochemicals, and atmospheric deposition [1, 2]. Heavy metals constitute an ill-defined group of inorganic chemical hazards, and those most commonly found at contaminated sites are lead (Pb), chromium (Cr), arsenic (As), zinc (Zn), cadmium (Cd), copper (Cu), mercury (Hg), and nickel (Ni) [3]. Soils are the major sink for heavy metals released into the environment by aforementioned anthropogenic activities and unlike organic contaminants which are oxidized to carbon (IV) oxide by microbial action, most metals do not undergo microbial or chemical degradation [4], and their total concentration in soils persists for a long time after their introduction [5]. Changes in their chemical forms (speciation) and bioavailability are, however, possible. The presence of toxic metals in soil can severely inhibit the biodegradation of organic contaminants [6]. Heavy metal contamination of soil may pose risks and hazards to humans and the ecosystem through: direct ingestion or contact with contaminated soil, the food chain (soil-plant-human or soil-plant-animal-human), drinking of contaminated ground water, reduction in food quality (safety and marketability) via phytotoxicity, reduction in land usability for agricultural production causing food insecurity, and land tenure problems [7–9].

The adequate protection and restoration of soil ecosystems contaminated by heavy metals require their characterization and remediation. Contemporary legislation respecting environmental protection and public health, at both national and international levels, are based on data that characterize chemical properties of environmental phenomena, especially those that reside in our food chain [10]. While soil characterization would provide an insight into heavy metal speciation and bioavailability, attempt at remediation of heavy metal contaminated soils would entail knowledge of the source of contamination, basic chemistry, and environmental and associated health effects (risks) of these heavy metals. Risk assessment is an effective scientific tool which enables decision makers to manage sites so contaminated in a cost-effective manner while preserving public and ecosystem health [11].

Immobilization, soil washing, and phytoremediation techniques are frequently listed among the best demonstrated available technologies (BDATs) for remediation of heavy metal-contaminated sites [3]. In spite of their cost-effectiveness and environment friendliness, field applications of these technologies have only been reported in developed countries. In most developing countries, these are yet to become commercially available technologies possibly due to the inadequate awareness of their inherent advantages and principles of operation. With greater awareness by the governments and the public of the implications of contaminated soils on human and animal health, there has been increasing interest amongst the scientific community in the development of technologies to remediate contaminated sites [12]. In developing countries with great population density and scarce funds available for environmental restoration, low-cost and ecologically sustainable remedial options are required to restore contaminated lands so as to reduce the associated risks, make the land resource available for agricultural production, enhance food security, and scale down land tenure problems.

In this paper, scattered literature is utilized to review the possible sources of contamination, basic chemistry, and the associated environmental and health risks of priority heavy metals (Pb, Cr, As, Zn, Cd, Cu, Hg, and Ni) which can provide insight into heavy metal speciation, bioavailability, and hence selection of appropriate remedial options. The principles, advantages, and disadvantages of immobilization, soil washing, and phytoremediation techniques as options for soil cleanup are also presented.

SOURCES OF HEAVY METALS IN CONTAMINATED SOILS

Heavy metals occur naturally in the soil environment from the pedogenetic processes of weathering of parent materials at levels that are regarded as trace ($<1000\,\mathrm{mg\,kg^{-1}}$) and rarely toxic [10, 13]. Due to the disturbance and acceleration of nature's slowly occurring geochemical cycle of metals by man, most soils of rural and urban environments may accumulate one or more of the heavy metals above defined background values high enough to cause risks to human health, plants, animals, ecosystems, or other media [14]. The heavy metals essentially become contaminants in the soil environments because (i) their rates of generation via man-made cycles are more rapid relative to natural ones, (ii) they become transferred from mines to random environmental locations where higher potentials of direct exposure occur, (iii) the concentrations of the metals in discarded products are relatively high compared to those in the receiving environment, and (iv) the chemical form (species) in which a metal is found in the receiving environmental system may render it more bioavailable [14]. A simple mass balance of the heavy metals in the soil can be expressed as follows [15, 16]:

$$M_{\text{total}} = \left(M_p + M_a + M_f + M_{\text{ag}} + M_{\text{ow}} + M_{\text{ip}}\right) - (M_{\text{cr}} + M_l), \tag{1}$$

where "M" is the heavy metal, "p" is the parent material, "a" is the atmospheric deposition, "f" is the fertilizer sources, "ag" are the agrochemical sources, "ow" are the organic waste sources, "ip" are other inorganic pollutants, "cr" is crop removal, and "l" is the losses by leaching, volatilization, and so forth. It is projected that the anthropogenic emission into the atmosphere, for several heavy metals, is one-to-three orders of magnitude higher than natural fluxes [17]. Heavy metals in the soil from anthropogenic sources tend to be more mobile, hence bioavailable than pedogenic, or lithogenic ones [18, 19]. Metal-bearing solids at contaminated sites can originate from a wide variety of anthropogenic sources in the form of metal mine tailings, disposal of high metal wastes in improperly protected landfills, leaded gasoline and lead-based paints, land application of fertilizer, animal manures, biosolids (sewage sludge), compost, pesticides, coal combustion residues, petrochemicals, and atmospheric deposition [1, 2, 20] are discussed hereunder.

Fertilizers

Historically, agriculture was the first major human influence on the soil [21]. To grow and complete the lifecycle, plants must acquire not only macronutrients (N, P, K, S, Ca, and Mg), but also essential micronutrients. Some soils are deficient in the heavy metals (such as Co, Cu, Fe, Mn, Mo, Ni, and Zn) that are essential for healthy plant growth [22], and crops may be supplied with these as an addition to the soil or as a foliar spray. Cereal crops grown on Cu-deficient soils are occasionally treated with Cu as an addition to the soil, and Mn may similarly be supplied to cereal and root crops. Large quantities of fertilizers are regularly added to soils in intensive farming systems to provide adequate N, P, and K for crop growth. The compounds used to supply these elements contain trace amounts of heavy metals (e.g., Cd and Pb) as impurities, which, after continued fertilizer, application may significantly increase their content in the soil [23]. Metals, such as Cd and Pb, have no known physiological activity. Application of certain phosphatic fertilizers inadvertently adds Cd and other potentially toxic elements to the soil, including F, Hg, and Pb [24].

Pesticides

Several common pesticides used fairly extensively in agriculture and horticulture in the past contained substantial concentrations of metals. For instance in the recent past, about 10% of the chemicals have approved for use as insecticides and fungicides in UK were based on compounds which contain Cu, Hg, Mn, Pb, or Zn. Examples of such pesticides are copper-containing fungicidal sprays such as Bordeaux mixture (copper sulphate) and copper oxychloride [23]. Lead arsenate was used in fruit orchards for many years to control some parasitic insects. Arsenic-containing compounds were also used extensively to control cattle ticks and to control pests in banana in New Zealand and Australia, timbers have been preserved with formulations of Cu, Cr, and As (CCA), and there are now many derelict sites where soil concentrations of these elements greatly exceed background concentrations. Such contamination has the potential to cause problems, particularly if sites are redeveloped for other agricultural or nonagricultural purposes. Compared with fertilizers, the use of such materials has been more localized, being restricted to particular sites or crops [8].

Biosolids and Manures

The application of numerous biosolids (e.g., livestock manures, composts, and municipal sewage sludge) to land inadvertently leads to the accumulation of heavy metals such as As, Cd, Cr, Cu, Pb, Hg, Ni, Se, Mo, Zn, Tl, Sb, and so forth, in the soil [20]. Certain animal wastes such as poultry, cattle, and pig manures produced in agriculture are commonly applied to crops and pastures either as solids or slurries [25]. Although most manures are seen as valuable fertilizers, in the pig and poultry industry, the Cu and Zn added to diets as growth promoters and As contained in poultry health products may also have the potential to cause metal contamination of the soil [25, 26]. The manures produced from animals on such diets contain high concentrations of As, Cu, and Zn and, if repeatedly applied to restricted areas of land, can cause considerable buildup of these metals in the soil in the long run.

Biosolids (sewage sludge) are primarily organic solid products, produced by wastewater treatment processes that can be beneficially recycled [27]. Land application of biosolids materials is a common practice in many countries that allow the reuse of biosolids produced by urban populations [28]. The term sewage sludge is used in many references because of its wide recognition and its regulatory definition. However, the term biosolids is becoming more common as a replacement for sewage sludge because it is thought to reflect more accurately the beneficial characteristics inherent to sewage sludge [29]. It is estimated that in the United States, more than half of approximately 5.6 million dry tonnes of sewage sludge used or disposed of annually is land applied, and agricultural utilization of biosolids occurs in every region of the country. In the European community, over 30% of the sewage sludge is used as fertilizer in agriculture [29]. In Australia over 175 000 tonnes of dry biosolids are produced each year by the major metropolitan authorities, and currently most biosolids applied to agricultural land are used in arable cropping situations where they can be incorporated into the soil [8].

There is also considerable interest in the potential for composting biosolids with other organic materials such as sawdust, straw, or garden waste. If this trend continues, there will be implications for metal contamination of soils. The potential of biosolids for contaminating soils with heavy metals has caused great concern about their application in agricultural practices [30]. Heavy metals most commonly found in biosolids are Pb, Ni, Cd, Cr, Cu, and Zn, and the metal concentrations are governed by the nature and the intensity of the industrial activity, as well as the type of process employed during the biosolids treatment [31]. Under certain conditions, metals added to soils in applications of biosolids can be leached downwards through the soil profile and can have the potential to contaminate groundwater [32]. Recent studies on some New Zealand soils treated with biosolids have shown increased concentrations of Cd, Ni, and Zn in drainage leachates [33, 34].

Wastewater

The application of municipal and industrial wastewater and related effluents to land dates back 400 years and now is a common practice in many parts of the world [35]. Worldwide, it is estimated that 20 million hectares of arable land are irrigated with waste water. In several Asian and African cities, studies suggest that agriculture based on wastewater irrigation accounts for 50 percent of the vegetable supply to urban areas [36]. Farmers generally are not bothered about environmental benefits or hazards and are primarily interested in maximizing their yields and profits. Although the metal concentrations in wastewater effluents are usually relatively low, long-term irrigation of land with such can eventually result in heavy metal accumulation in the soil.

Metal Mining and Milling Processes and Industrial Wastes

Mining and milling of metal ores coupled with industries have bequeathed many countries, the legacy of wide distribution of metal contaminants in soil. During mining, tailings (heavier and larger particles settled at the bottom of the flotation cell during mining) are directly discharged into natural depressions, including onsite wetlands resulting in elevated concentrations [37]. Extensive Pb and zinc Zn ore mining and smelting have resulted in contamination of soil that poses risk to human and ecological health. Many reclamation methods used for these sites are lengthy and expensive and may not restore soil productivity. Soil heavy metal environmental risk to humans is related to bioavailability. Assimilation pathways include the ingestion of plant material grown in (food chain), or the direct ingestion (oral bioavailability) of, contaminated soil [38].

Other materials are generated by a variety of industries such as textile, tanning, petrochemicals from accidental oil spills or utilization of petroleum-based products, pesticides, and pharmaceutical facilities and are highly variable in composition. Although some are disposed of on land, few have benefits to agriculture or forestry. In addition, many are potentially hazardous because of their contents of heavy metals (Cr, Pb, and Zn) or toxic organic compounds and are seldom, if ever, applied to land. Others are very low in plant nutrients or have no soil conditioning properties [25].

Air-Borne Sources

Airborne sources of metals include stack or duct emissions of air, gas, or vapor streams, and fugitive emissions such as dust from storage areas or waste piles. Metals from airborne sources are generally released as particulates contained in the gas stream. Some metals such as As, Cd, and Pb can also volatilize during high-temperature processing. These metals will convert to oxides and condense as fine particulates unless a reducing atmosphere is maintained [39]. Stack emissions can be distributed over a wide area by natural air currents until dry and/or wet precipitation mechanisms remove them from the gas stream. Fugitive emissions are often distributed over a much smaller area because emissions are made near the ground. In general, contaminant concentrations are lower in fugitive emissions compared to stack emissions. The type and concentration of metals emitted from both types of sources will depend on site-specific conditions. All solid particles in smoke from fires and in other emissions from factory chimneys are eventually deposited on land or sea; most forms of fossil fuels contain some heavy metals and this is, therefore, a form of contamination which has been continuing on a large scale since the industrial revolution began. For example, very high concentration of Cd, Pb, and Zn has been found in plants and soils adjacent to smelting works. Another major source of soil contamination is the aerial emission of Pb from the combustion of petrol containing tetraethyl lead; this contributes substantially to the content of Pb in soils in urban areas and in those adjacent to major roads. Zn and Cd may also be added to soils adjacent to roads, the sources being tyres, and lubricant oils [40].

BASIC SOIL CHEMISTRY AND POTENTIAL RISKS OF HEAVY METALS

The most common heavy metals found at contaminated sites, in order of abundance are Pb, Cr, As, Zn, Cd, Cu, and Hg [40]. Those metals are important since they are capable of decreasing crop production due to the risk of bioaccumulation and biomagnification in the food chain. There's also the risk of superficial and groundwater contamination. Knowledge of the basic chemistry, environmental, and associated health effects of these heavy metals is necessary in understanding their speciation, bioavailability, and remedial options. The fate and transport of a heavy metal in soil depends significantly on the chemical form and speciation of the metal. Once in the soil, heavy metals are adsorbed by initial fast reactions (minutes, hours), followed by slow adsorption reactions (days, years) and are, therefore, redistributed into different chemical forms with varying bioavailability, mobility, and toxicity [41, 42]. This distribution is believed to be controlled by reactions of heavy metals in soils such as (i) mineral precipitation and dissolution, (ii) ion exchange, adsorption, and desorption, (iii) aqueous complexation, (iv) biological immobilization and mobilization, and (v) plant uptake [43].

Lead

Lead is a metal belonging to group IV and period 6 of the periodic table with atomic number 82, atomic mass 207.2, density $11.4 \, g \, cm^{-3}$, melting point 327.4°C, and boiling point 1725°C. It is a naturally occurring, bluish-gray metal usually found as a mineral combined with other elements, such as sulphur (i.e., PbS, $PbSO_4$), or oxygen ($PbCO_3$), and ranges from 10 to 30 $mg \, kg^{-1}$ in the earth's crust [44]. Typical mean Pb concentration for surface soils worldwide averages 32 $mg \, kg^{-1}$ and ranges from 10 to 67 mg kg^{-1} [10]. Lead ranks fifth behind Fe, Cu, Al, and Zn in industrial production of metals. About half of the Pb used in the U.S. goes for the manufacture of Pb storage batteries. Other uses include solders, bearings, cable covers, ammunition, plumbing, pigments, and caulking. Metals commonly alloyed with Pb are antimony (in storage batteries), calcium (Ca) and tin (Sn) (in maintenance-free storage batteries), silver (Ag) (for solder and anodes), strontium (Sr) and Sn (as anodes in electrowinning processes), tellurium (Te) (pipe and sheet in chemical installations and nuclear shielding), Sn (solders), and antimony (Sb), and Sn (sleeve bearings, printing, and high-detail castings) [45].

Ionic lead, Pb(II), lead oxides and hydroxides, and lead-metal oxyanion complexes are the general forms of Pb that are released into the soil, groundwater, and surface waters. The most stable forms of lead are Pb(II) and lead-hydroxy complexes. Lead(II) is the most common and reactive form of Pb, forming mononuclear and polynuclear oxides and hydroxides [3]. The predominant insoluble Pb compounds are lead phosphates, lead carbonates (form when the pH is above 6), and lead (hydr)oxides [46]. Lead sulfide (PbS) is the most stable solid form within the soil matrix and forms under reducing conditions, when increased concentrations of sulfide are present. Under anaerobic conditions a volatile organolead (tetramethyl lead) can be formed due to microbial alkylation [3].

Lead(II) compounds are predominantly ionic (e.g., Pb^{2+} SO_4^{2-}), whereas Pb(IV) compounds tend to be covalent (e.g., tetraethyl lead, $Pb(C_2H_5)_4$). Some Pb (IV) compounds, such as PbO_2, are strong oxidants. Lead forms several basic salts, such as $Pb(OH)_2 \cdot 2PbCO_3$, which was once the most widely used white paint pigment and the source of considerable chronic lead poisoning to children who ate peeling white paint. Many compounds of Pb(II) and a few Pb(IV) compounds are useful. The two most common of these are lead dioxide and lead sulphate, which are participants in the reversible reaction that occurs during the charge and discharge of lead storage battery.

In addition to the inorganic compounds of lead, there are a number of organolead compounds such as tetraethyl lead. The toxicities and environmental effects of organolead compounds are particularly noteworthy because of the former widespread use and distribution of tetraethyllead as a gasoline additive. Although more than 1000 organolead compounds have been synthesized, those of commercial and toxicological importance are largely limited to the alkyl (methyl and ethyl) lead compounds and their salts (e.g., dimethyldiethyllead, trimethyllead chloride, and diethyllead dichloride).

Inhalation and ingestion are the two routes of exposure, and the effects from both are the same. Pb accumulates in the body organs (i.e., brain), which may lead to poisoning (plumbism) or even death. The gastrointestinal tract, kidneys, and central nervous system are also affected by the presence of lead. Children exposed to lead are at risk for impaired development, lower IQ, shortened attention span, hyperactivity, and mental deterioration, with children under the age of six being at a more substantial risk. Adults usually experience decreased reaction time, loss of memory, nausea, insomnia, anorexia, and weakness of the joints when exposed to lead [47]. Lead is not an essential element. It is well known to be toxic and its effects have been more extensively reviewed than the effects of other trace metals. Lead can cause serious injury to the brain, nervous system, red blood cells, and kidneys [48]. Exposure to lead can result in a wide range of biological effects depending on the level and duration of exposure. Various effects occur over a broad range of doses, with the developing young and infants being more sensitive than adults. Lead poisoning, which is so severe as to cause evident illness, is now very rare. Lead performs no known essential function in the human body, it can merely do harm after uptake from food, air, or water. Lead is a particularly dangerous chemical, as it can accumulate in individual organisms, but also in entire food chains.

The most serious source of exposure to soil lead is through direct ingestion (eating) of contaminated soil or dust. In general, plants do not absorb or accumulate lead. However, in soils testing high in lead, it is possible for some lead to be taken up. Studies have shown that lead does not readily accumulate in the fruiting parts of vegetable and fruit crops (e.g., corn, beans, squash, tomatoes, strawberries, and apples). Higher concentrations are more likely to be found in leafy vegetables (e.g., lettuce) and on the surface of root crops (e.g., carrots). Since plants do not take up large quantities of soil lead, the lead levels in soil considered safe for plants will be much higher than soil lead levels where eating of soil is a concern (pica). Generally, it has been considered safe to use garden produce grown in soils with total lead levels less than 300 ppm. The risk of lead poisoning through the food chain increases as the soil lead level rises above this concentration. Even at soil levels above 300 ppm, most of the risk is from lead contaminated soil or dust deposits on the plants rather than from uptake of lead by the plant [49].

Chromium

Chromium is a first-row d-block transition metal of group VIB in the periodic table with the following properties: atomic number 24, atomic mass 52, density $7.19\,g\,cm^{-3}$, melting point 1875°C, and boiling point 2665°C. It is one of the less common elements and does not occur naturally in elemental form, but only in compounds. Chromium is mined as a primary ore product in the form of the mineral chromite, $FeCr_2O_4$. Major sources of Cr-contamination include releases from electroplating processes and the disposal of Cr containing wastes [39]. Chromium(VI) is the form of Cr commonly found at contaminated sites. Chromium can also occur in the +III oxidation state, depending on pH and redox conditions. Chromium(VI) is the dominant form of Cr in shallow aquifers where aerobic conditions exist. Chromium(VI) can be reduced to Cr(III) by soil organic matter, S^{2-} and Fe^{2+} ions under anaerobic conditions often encountered in deeper groundwater. Major Cr(VI) species include chromate (CrO_4^{2-}) and dichromate ($Cr_2O_7^{2-}$) which precipitate readily in the presence of metal cations (especially Ba^{2+}, Pb^{2+}, and Ag^+). Chromate and dichromate also adsorb on soil surfaces, especially iron and aluminum oxides. Chromium(III) is the dominant form of Cr at low pH (<4). Cr^{3+} forms solution complexes with NH_3, OH^-, Cl^-, F^-, CN^-, SO_4^{2-}, and soluble organic ligands. Chromium(VI) is the more toxic form of chromium and is also more mobile. Chromium(III) mobility is decreased by adsorption to clays and oxide minerals below pH 5 and low solubility above pH 5 due to the formation of $Cr(OH)_3(s)$ [50]. Chromium mobility depends on sorption characteristics of the soil, including clay content, iron oxide content, and the amount of organic matter present. Chromium can be transported by surface runoff to surface waters in its soluble or precipitated form. Soluble and un-adsorbed chromium complexes can leach from soil into groundwater. The leachability of Cr(VI) increases as soil pH increases. Most of Cr released into natural waters is particle associated, however, and is ultimately deposited into the sediment [39]. Chromium is associated with allergic dermatitis in humans [21].

Arsenic

Arsenic is a metalloid in group VA and period 4 of the periodic table that occurs in a wide variety of minerals, mainly as As_2O_3, and can be recovered from processing of ores containing mostly Cu, Pb, Zn, Ag and Au. It is also present in ashes from coal combustion. Arsenic has the following properties: atomic number 33, atomic mass 75, density $5.72\,g\,cm^{-3}$, melting point 817°C, and boiling point 613°C, and exhibits fairly complex chemistry and can be present in several oxidation states (−III, 0, III, V) [39]. In aerobic environments, As (V) is dominant, usually in the form of arsenate (AsO_4^{3-}) in various protonation states: H_3AsO_4, $H_2AsO_4^-$, $HAsO_4^{2-}$, and AsO_4^{3-}. Arsenate and other anionic forms of arsenic behave as chelates and can precipitate when metal cations are present [51]. Metal arsenate complexes are stable only under certain conditions. Arsenic (V) can also coprecipitate with or adsorb onto iron oxyhydroxides under acidic and moderately reducing conditions. Coprecipitates are immobile under these conditions, but arsenic mobility increases as pH increases [39]. Under reducing conditions As(III) dominates, existing as arsenite (AsO_3^{3-}), and its protonated forms H_3AsO_3, $H_2AsO_3^-$, and $HAsO_3^{2-}$. Arsenite can adsorb or coprecipitate with metal sulfides and has a high affinity for other sulfur compounds. Elemental arsenic and arsine, AsH_3, may be present under extreme reducing conditions. Biotransformation (via methylation) of arsenic creates methylated derivatives of arsine, such as dimethyl arsine $HAs(CH_3)_2$ and trimethylarsine $As(CH_3)_3$ which are highly volatile. Since arsenic is often present in anionic form, it does not form complexes with simple anions such as Cl^- and SO_4^{2-}. Arsenic speciation also includes organometallic forms such as methylarsinic acid $(CH_3)AsO_2H_2$ and dimethylarsinic acid $(CH_3)_2AsO_2H$. Many As compounds adsorb strongly to soils and are therefore transported only over short distances in groundwater and surface water. Arsenic is associated with skin damage, increased risk of cancer, and problems with circulatory system [21].

Zinc

Zinc is a transition metal with the following characteristics: period 4, group IIB, atomic number 30, atomic mass 65.4, density $7.14\,g\,cm^{-3}$, melting point 419.5°C, and boiling point 906°C. Zinc occurs naturally in soil (about 70 mg kg^{-1} in crustal rocks) [52], but Zn concentrations are rising unnaturally, due to anthropogenic additions. Most Zn is added during industrial activities, such as mining, coal, and waste combustion and steel processing. Many foodstuffs contain certain concentrations of Zn. Drinking water also contains certain amounts of Zn, which may be higher when it is stored in metal tanks. Industrial sources or toxic waste sites may cause the concentrations of Zn in drinking water to reach levels that can cause health problems. Zinc is a trace element that is essential for human health. Zinc shortages can cause birth defects. The world's Zn production is still on the rise which means that more and more Zn ends up in the environment. Water is polluted with Zn, due to the presence of large quantities present in the wastewater of industrial plants. A consequence is that Zn-polluted sludge is continually being deposited by rivers on their banks. Zinc may also increase the acidity of waters. Some fish can accumulate Zn in their bodies, when they live in Zn-contaminated waterways. When Zn enters the bodies of these fish, it is able to biomagnify up the food chain. Water-soluble zinc that is located in soils can contaminate groundwater. Plants often have a Zn uptake that their systems cannot handle, due to the accumulation of Zn in soils. Finally, Zn can interrupt the activity in soils, as it negatively influences the activity of microorganisms and earthworms, thus retarding the breakdown of organic matter [53].

Cadmium

Cadmium is located at the end of the second row of transition elements with atomic number 48, atomic weight 112.4, density 8.65 g cm^{-3}, melting point 320.9°C, and boiling point 765°C. Together with Hg and Pb, Cd is one of the big three heavy metal poisons and is not known for any essential biological function. In its compounds, Cd occurs as the divalent Cd(II) ion. Cadmium is directly below Zn in the periodic table and has a chemical similarity to that of Zn, an essential micronutrient for plants and animals. This may account in part for Cd's toxicity; because Zn being an essential trace element, its substitution by Cd may cause the malfunctioning of metabolic processes [54].

The most significant use of Cd is in Ni/Cd batteries, as rechargeable or secondary power sources exhibiting high output, long life, low maintenance, and high tolerance to physical and electrical stress.

Cadmium coatings provide good corrosion resistance coating to vessels and other vehicles, particularly in high-stress environments such as marine and aerospace. Other uses of cadmium are as pigments, stabilizers for polyvinyl chloride (PVC), in alloys and electronic compounds. Cadmium is also present as an impurity in several products, including phosphate fertilizers, detergents and refined petroleum products. In addition, acid rain and the resulting acidification of soils and surface waters have increased the geochemical mobility of Cd, and as a result its surface-water concentrations tend to increase as lake water pH decreases [54]. Cadmium is produced as an inevitable byproduct of Zn and occasionally lead refining. The application of agricultural inputs such as fertilizers, pesticides, and biosolids (sewage sludge), the disposal of industrial wastes or the deposition of atmospheric contaminants increases the total concentration of Cd in soils, and the bioavailability of this Cd determines whether plant Cd uptake occurs to a significant degree [28]. Cadmium is very biopersistent but has few toxicological properties and, once absorbed by an organism, remains resident for many years.

Since the 1970s, there has been sustained interest in possible exposure of humans to Cd through their food chain, for example, through the consumption of certain species of shellfish or vegetables. Concern regarding this latter route (agricultural crops) led to research on the possible consequences of applying sewage sludge (Cd-rich biosolids) to soils used for crops meant for human consumption, or of using cadmium-enriched phosphate fertilizer [54]. This research has led to the stipulation of highest permissible concentrations for a number of food crops [8].

Cadmium in the body is known to affect several enzymes. It is believed that the renal damage that results in proteinuria is the result of Cd adversely affecting enzymes responsible for reabsorption of proteins in kidney tubules. Cadmium also reduces the activity of delta-aminolevulinic acid synthetase, arylsulfatase, alcohol dehydrogenase, and lipoamide dehydrogenase, whereas it enhances the activity of delta-aminolevulinic acid dehydratase, pyruvate dehydrogenase, and pyruvate decarboxylase [45]. The most spectacular and publicized occurrence of cadmium poisoning resulted from dietary intake of cadmium by people in the Jintsu River Valley, near Fuchu, Japan. The victims were afflicted by itai itai disease, which means ouch, ouch in Japanese. The symptoms are the result of painful osteomalacia (bone disease) combined with kidney malfunction. Cadmium poisoning in the Jintsu River Valley was attributed to irrigated rice contaminated from an upstream mine producing Pb, Zn, and Cd. The major threat to human health is chronic accumulation in the kidneys leading to kidney dysfunction. Food intake and tobacco smoking are the main routes by which Cd enters the body [45].

Copper

Copper is a transition metal which belongs to period 4 and group IB of the periodic table with atomic number 29, atomic weight 63.5, density 8.96 g cm^{-3}, melting point 1083°C and boiling point 2595°C. The metal's average density and concentrations in crustal rocks are 8.1 × 10^3 kg m^{-3} and 55 mg kg^{-1}, respectively [52].

Copper is the third most used metal in the world [55]. Copper is an essential micronutrient required in the growth of both plants and animals. In humans, it helps in the production of blood haemoglobin. In plants, Cu is especially important in seed production, disease resistance, and regulation of water. Copper is indeed essential, but in high doses it can cause anaemia, liver and kidney damage, and stomach and intestinal irritation. Copper normally occurs in drinking water from Cu pipes, as well as from additives designed to control algal growth. While Cu's interaction with the environment is complex, research shows that most Cu introduced into the environment is, or rapidly becomes, stable and results in a form which does not pose a risk to the environment. In fact, unlike some man-made materials, Cu is not magnified in the body or bioaccumulated in the food chain. In the soil, Cu strongly complexes to the organic implying that only a small fraction of copper will be found in solution as ionic copper, Cu(II). The solubility of Cu is drastically increased at pH 5.5 [56], which is rather close to the ideal farmland pH of 6.0–6.5 [57].

Copper and Zn are two important essential elements for plants, microorganisms, animals, and humans. The connection between soil and water contamination and metal uptake by plants is determined by many chemical and physical soil factors as well as the physiological properties of the crops. Soils contaminated with trace metals may pose both direct and indirect threats: direct, through negative effects of metals on crop growth and yield, and indirect, by entering the human food chain with a potentially negative impact on human health. Even a reduction of crop yield by a few percent could lead to a significant long-term loss in production and income. Some food importers are now specifying acceptable maximum contents of metals in food, which might limit the possibility for the farmers to export their contaminated crops [36].

Mercury

Mercury belongs to same group of the periodic table with Zn and Cd. It is the only liquid metal at stp. It has atomic number 80, atomic weight 200.6, density $13.6\,g\,cm^{-3}$, melting point $-13.6°C$, and boiling point $357°C$ and is usually recovered as a byproduct of ore processing [39]. Release of Hg from coal combustion is a major source of Hg contamination. Releases from manometers at pressure-measuring stations along gas/oil pipelines also contribute to Hg contamination. After release to the environment, Hg usually exists in mercuric (Hg^{2+}), mercurous (Hg_2^{2+}), elemental (Hg^0), or alkylated form (methyl/ethyl mercury). The redox potential and pH of the system determine the stable forms of Hg that will be present. Mercurous and mercuric mercury are more stable under oxidizing conditions. When mildly reducing conditions exist, organic or inorganic Hg may be reduced to elemental Hg, which may then be converted to alkylated forms by biotic or abiotic processes. Mercury is most toxic in its alkylated forms which are soluble in water and volatile in air [39]. Mercury(II) forms strong complexes with a variety of both inorganic and organic ligands, making it very soluble in oxidized aquatic systems [51]. Sorption to soils, sediments, and humic materials is an important mechanism for the removal of Hg from solution. Sorption is pH dependent and increases as pH increases. Mercury may also be removed from solution by coprecipitation with sulphides. Under anaerobic conditions, both organic and inorganic forms of Hg may be converted to alkylated forms by microbial activity, such as by sulfur-reducing bacteria. Elemental mercury may also be formed under anaerobic conditions by demethylation of methyl mercury, or by reduction of Hg(II). Acidic conditions (pH < 4) also favor the formation of methyl mercury, whereas higher pH values favor precipitation of HgS(s) [39]. Mercury is associated with kidney damage [21].

Nickel

Nickel is a transition element with atomic number 28 and atomic weight 58.69. In low pH regions, the metal exists in the form of the nickelous ion, Ni(II). In neutral to slightly alkaline solutions, it precipitates as nickelous hydroxide, $Ni(OH)_2$, which is a stable compound. This precipitate readily dissolves in acid solutions forming Ni(III) and in very alkaline conditions; it forms nickelite ion, $HNiO_2$, that is soluble in water. In very oxidizing and alkaline conditions, nickel exists in form of the stable nickelo-nickelic oxide, Ni_3O_4, that is soluble in acid solutions. Other nickel oxides such as nickelic oxide, Ni_2O_3, and nickel peroxide, NiO_2, are unstable in alkaline solutions and decompose by giving off oxygen. In acidic regions, however, these solids dissolve producing Ni^{2+} [58].

Nickel is an element that occurs in the environment only at very low levels and is essential in small doses, but it can be dangerous when the maximum tolerable amounts are exceeded. This can cause various kinds of cancer on different sites within the bodies of animals, mainly of those that live near refineries. The most common application of Ni is an ingredient of steel and other metal products. The major sources of nickel contamination in the soil are metal plating industries, combustion of fossil fuels, and nickel mining and electroplating [59]. It is released into the air by power plants and trash incinerators and settles to the ground after undergoing precipitation reactions. It usually takes a long time for nickel to be removed from air. Nickel can also end up in surface water when it is a part of wastewater streams. The larger part of all Ni compounds that are released to the environment will adsorb to sediment or soil particles and become immobile as a result. In acidic soils, however, Ni becomes more mobile and often leaches down to the adjacent groundwater. Microorganisms can also suffer from growth decline due to the presence of Ni, but they usually develop resistance to Ni after a while. Nickel is not known to accumulate in plants or animals and as a result Ni has not been found to biomagnify up the food chain. For animals Ni is an essential foodstuff in small amounts. The primary source of mercury is the sulphide ore cinnabar.

SOIL CONCENTRATION RANGES AND REGULATORY GUIDELINES FOR SOME HEAVY METALS

The specific type of metal contamination found in a contaminated soil is directly related to the operation that occurred at the site. The range of contaminant concentrations and the physical and chemical forms of contaminants will also depend on activities and disposal patterns for contaminated wastes on the site. Other factors that may influence the form, concentration, and distribution of metal contaminants include soil and ground-water chemistry and local transport mechanisms [3].

Soils may contain metals in the solid, gaseous, or liquid phases, and this may complicate analysis and interpretation of reported results. For example, the most common method for determining the concentration of metals contaminants in soil is via total elemental analysis (USEPA Method 3050). The level of metal contamination determined by this method is expressed as mg metal kg^{-1} soil. This analysis does not specify requirements for the moisture content of the soil and may therefore include soil water. This measurement may also be reported on a dry soil basis. The level of contamination may also be reported as leachable metals as determined by leach tests, such as the toxicity characteristic leaching procedure (TCLP) (USEPA Method 1311) or the synthetic precipitation-leaching procedure, or SPLP test (USEPA Method 1312). These procedures measure the concentration of metals in leachate from soil contacted with an acetic acid solution (TCLP) [60] or a dilute solution of sulfuric and nitric acid (SPLP). In this case, metal contamination is expressed in mgL^{-1} of the leachable metal. Other types of leaching tests have been proposed including sequential extraction procedures [61, 62] and extraction of acid volatile sulfide [63]. Sequential procedures contact the solid with a series of extractant solutions that are designed to dissolve different fractions of the associated metal. These tests may provide insight into the different forms of metal contamination present. Contaminant concentrations can be measured directly in metals-contaminated water. These concentrations are most commonly expressed as total dissolved metals in mass concentrations (mg L^{-1} or gL^{-1}) or in molar concentrations (mol L^{-1}). In dilute solutions, a mg L^{-1} is equivalent to one part per million (ppm), and a gL^{-1} is equivalent to one part per billion (ppb).

Riley et al. [64] and NJDEP [65] have reported soil concentration ranges and regulatory guidelines for some heavy metals (Table 1). In Nigeria, in the interim period, whilst suitable parameters are being developed, the Department of Petroleum Resources [60] has recommended guidelines on remediation of contaminated land based on two parameters intervention values and target values (Table 2).

Table 1: Soil concentration ranges and regulatory guidelines for some heavy metals

Metal	Soil concentration range† (mg kg−1)	Regulatory limits‡ (mg kg−1)
Pb	1.00–69 000	600
Cd	0.10–345	100
Cr	0.05–3 950	100
Hg	<0.01–1 800	270
Zn	150–5 000	1 500

†[64]; ‡Nonresidential direct contact soil clean-up criteria [65].

Table 2: Target and intervention values for some metals for a standard soil [60]

Metal	Target value (mg kg−1)	Intervention value (mg kg−1)
Ni	140.00	720.00
Cu	0.30	10.00
Zn	—	—
Cd	100.00	380.00
Pb	35.00	210.00
As	200	625
Cr	20	240
Hg	85	530

The intervention values indicate the quality for which the functionality of soil for human, animal, and plant life are, or threatened with being seriously impaired. Concentrations in excess of the intervention values correspond to serious contamination. Target values indicate the soil quality required for sustainability or expressed in terms of remedial policy, the soil quality required for the full restoration of the soil's functionality for human, animal, and plant life. The target values therefore indicate the soil quality levels ultimately aimed at.

REMEDIATION OF HEAVY METAL-CONTAMINATED SOILS

The overall objective of any soil remediation approach is to create a final solution that is protective of human health and the environment [66]. Remediation is generally subject to an array of regulatory requirements and can also be based on assessments of human health and ecological risks where no legislated standards exist or where standards are advisory. The regulatory authorities will normally accept remediation strategies that centre on reducing metal bioavailability only if reduced bioavailability is equated with reduced risk, and if the bioavailability reductions are demonstrated to be long term [66]. For heavy metal-contaminated soils, the physical and chemical form of the heavy metal contaminant in soil strongly influences the selection of the appropriate remediation treatment approach. Information about the physical characteristics of the site and the type and level of contamination at the site must be obtained to enable accurate assessment of site contamination and remedial alternatives. The contamination in the soil should be characterized to establish the type, amount, and distribution of heavy metals in the soil. Once the site has been characterized, the desired level of each metal in soil must be determined. This is done by comparison of observed heavy metal concentrations with soil quality standards for a particular regulatory domain, or by performance of a site-specific risk assessment. Remediation goals for heavy metals may be set as total metal concentration or as leachable metal in soil, or as some combination of these.

Several technologies exist for the remediation of metal-contaminated soil. Gupta et al. [67] have classified remediation technologies of contaminated soils into three categories of hazard-alleviating measures: (i) gentle in situ remediation, (ii) in situ harsh soil restrictive measures, and (iii) in situ or ex situ harsh soil destructive measures. The goal of the last two harsh alleviating measures is to avert hazards either to man, plant, or animal while the main goal of gentle in situ remediation is to restore the malfunctionality of soil (soil fertility), which allows a safe use of the soil. At present, a variety of approaches have been suggested for remediating contaminated soils. USEPA [68] has broadly classified remediation technologies for contaminated soils into (i) source control and (ii) containment remedies. Source control involves in situ and ex situ treatment technologies for sources of contamination. In situ or in place means that the contaminated soil is treated in its original place; unmoved, unexcavated; remaining at the site or in the subsurface. In situ treatment technologies treat or remove the contaminant from soil without excavation or removal of the soil. Ex situ means that the contaminated soil is moved, excavated, or removed from the site or subsurface. Implementation of ex situremedies requires excavation or removal of the contaminated soil. Containment remedies involve the construction of vertical engineered barriers (VEB), caps, and liners used to prevent the migration of contaminants.

Another classification places remediation technologies for heavy metal-contaminated soils under five categories of general approaches to remediation (Table 3): isolation, immobilization, toxicity reduction, physical separation, and extraction [3]. In practice, it may be more convenient to employ a hybrid of two or more of these approaches for more cost effectiveness. The key factors that may influence the applicability and selection of any of the available remediation technologies are: (i) cost, (ii) long-term effectiveness/permanence, (iii) commercial availability, (iv) general acceptance, (v) applicability to high metal concentrations, (vi) applicability to mixed wastes (heavy metals and organics), (vii) toxicity reduction, (viii) mobility reduction, and (ix) volume reduction. The present paper focuses on soil washing, phytoremediation, and immobilization techniques since they are among the best demonstrated available technologies (BDATs) for heavy metal-contaminated sites.

Table 3: Technologies for remediation of heavy metal-contaminated soils

Category	Remediation technologies
Isolation	(i) Capping (ii) subsurface barriers.
Immobilization	(i) Solidification/stabilization (ii) vitrification (iii) chemical treatment.
Toxicity and/ or mobility reduction	(i) Chemical treatment (ii) permeable treatment walls (iii) biological treatment bioaccumulation, phytoremediation (phytoextraction, phytostabilization, and rhizofiltration), bioleaching, biochemical processes.
Physical separation	
Extraction	(i) Soil washing, pyrometallurgical extraction, in situ soil flushing, and electrokinetic treatment.

Immobilization Techniques

Ex situ and in situ immobilization techniques are practical approaches to remediation of metal-contaminated soils. The ex situ technique is applied in areas where highly contaminated soil must be removed from its place of origin, and its storage is connected with a high ecological risk (e.g., in the case of radio nuclides). The method's advantages are: (i) fast and easy applicability and (ii) relatively low costs of investment and operation. The method's disadvantages include (i) high invasivity to the environment, (ii) generation of a significant amount of solid wastes (twice as large as volume after processing), (iii) the byproduct must be stored on a special landfill site, (iv) in the case of changing of the physicochemical condition in the side product or its surroundings, there is serious danger of the release of additional contaminants to the environment, and (v) permanent control of the stored wastes is required. In the in situ technique, the fixing agents amendments are applied on the unexcavated soil. The technique's advantages are (i) its low invasivity, (ii) simplicity and rapidity, (iii) relatively inexpensive, and (iv) small amount of wastes are produced, (v) high public acceptability, (vi) covers a broad spectrum of inorganic pollutants. The disadvantages of in situ immobilization are (i) its only a temporary solution (contaminants are still in the environment), (ii) the activation of pollutants may occur when soil physicochemical properties change, (iii) the reclamation process is applied only to the surface layer of soil (30–50 cm), and (iv) permanent monitoring is necessary [66, 69].

Immobilization technology often uses organic and inorganic amendment to accelerate the attenuation of metal mobility and toxicity in soils. The primary role of immobilizing amendments is to alter the original soil metals to more geochemically stable phases via sorption, precipitation, and complexation processes [70]. The mostly applied amendments include clay, cement, zeolites, minerals, phosphates, organic composts, and microbes [3,71]. Recent studies have indicated the potential of low-cost industrial residues such as red mud [72, 73] andtermitaria [74] in immobilization of heavy metals in contaminated soils. Due to the complexity of soil matrix and the limitations of current analytical techniques, the exact immobilization mechanisms have not been clarified, which could include precipitation, chemical adsorption and ion exchange, surface precipitation, formation of stable complexes with organic ligands, and redox reaction [75]. Most immobilization technologies can be performed ex situ or in situ. In situ processes are preferred due to the lower labour and energy requirements, but implementation of in situ will depend on specific site conditions.

Solidification/Stabilization (S/S)

Solidification involves the addition of binding agents to a contaminated material to impart physical/dimensional stability to contain contaminants in a solid product and reduce access by external agents through a combination of chemical reaction, encapsulation, and reduced permeability/surface area. Stabilization (also referred to as fixation) involves the addition of reagents to the contaminated soil to produce more chemically stable constituents. Conventional S/S is an established remediation technology for contaminated soils and treatment technology for hazardous wastes in many countries in the world [76].

The general approach for solidification/stabilization treatment processes involves mixing or injecting treatment agents to the contaminated soils. Inorganic binders (Table 4), such as clay (bentonite and kaolinite), cement, fly ash, blast furnace slag, calcium carbonate,

Fe/Mn oxides, charcoal, zeolite [9, 77], and organic stabilizers (Table 5) such as bitumen, composts, and manures [78], or a combination of organic-inorganic amendments may be used. The dominant mechanism by which metals are immobilized is by precipitation of hydroxides within the solid matrix [79, 80]. Solidification/stabilization technologies are not useful for some forms of metal contamination, such as species that exist as oxyanions (e.g., $Cr_2O_7^{2-}$, AsO_3^{-}) or metals that do not have low-solubility hydroxides (e.g., Hg). Solidification/stabilization may not be applicable at sites containing wastes that include organic forms of contamination, especially if volatile organics are present. Mixing and heating associated with binder hydration may release organic vapors. Pretreatment, such as air stripping or incineration, may be used to remove the organics and prepare the waste for metal stabilization/ solidification [39]. The application of S/S technologies will also be affected by the chemical composition of the contaminated matrix, the amount of water present, and the ambient temperature. These factors can interfere with the solidification/stabilization process by inhibiting bonding of the waste to the binding material, retarding the setting of the mixtures, decreasing the stability of the matrix, or reducing the strength of the solidified area [81].

Table 4: Organic amendments for heavy metal immobilization [82]

Material	Heavy metal immobilized
Bark saw dust (from timber industry)	Cd, Pb, Hg, Cu
Xylogen (from paper mill wastewater)	Zn, Pb, Hg
Chitosan (from crab meat canning industry)	Cd, Cr, Hg
Bagasse (from sugar cane)	Pb
Poultry manure (from poultry farm)	Cu, Pb, Zn, Cd
Cattle manure (from cattle farm)	Cd
Rice hulls (from rice processing)	Cd, Cr, Pb
Sewage sludge	Cd
Leaves	Cr, Cd
Straw	Cd, Cr, Pb

Table 5: Inorganic amendments for heavy metal immobilization [82]

Material	Heavy metal immobilized
Lime (from lime factory)	Cd, Cu, Ni, Pb, Zn,
Phosphate salt (from fertilizer plant)	Pb, Zn, Cu, Cd
Hydroxyapatite (from phosphorite)	Zn, Pb, Cu, Cd
Fly ash (from thermal power plant)	Cd, Pb, Cu, Zn, Cr
Slag (from thermal power plant)	Cd, Pb, Zn, Cr
Ca-montmorillonite (mineral)	Zn, Pb
Portland cement (from cement plant)	Cr, Cu, Zn, Pb
Bentonite	Pb

Cement-based binders and stabilizers are common materials used for implementation of S/S technologies [83]. Portland cement, a mixture of Ca silicates, aluminates, aluminoferrites, and sulfates, is an important cement-based material. Pozzolanic materials, which consist of small spherical particles formed by coal combustion (such as fly ash) and in lime and cement kilns, are also commonly used for S/S. Pozzolans exhibit cement-like properties, especially if the silica content is high. Portland cement and pozzolans can be used alone or together to obtain optimal properties for a particular site [84]. Organic binders may also be used to treat metals through polymer microencapsulation. This process uses organic materials such as bitumen, polyethylene, paraffins, waxes, and other polyolefins as thermoplastic or thermosetting resins. For polymer encapsulation, the organic materials are heated and mixed with the contaminated matrix at elevated temperatures (120° to 200°C). The organic materials polymerize and agglomerate the waste, and the waste matrix is encapsulated [84]. Organics are volatilized and collected, and the treated material is extruded for disposal or possible reuse (e.g., as paving material) [39]. The contaminated material may require pretreatment to separate rocks and debris and dry the feed material. Polymer encapsulation requires more energy and more complex equipment than cement-based S/S operations. Bitumen (asphalt) is the cheapest and most common thermoplastic binder [84]. Solidification/stabilization is achieved by mixing the contaminated material with appropriate amounts of binder/stabilizer and water. The mixture sets and cures to form a solidified matrix and contain the waste. The cure time and pour characteristics of the mixture and the final properties of the hardened cement depend upon the composition (amount of cement, pozzolan, and water) of the binder/stabilizer.

Ex situ S/S can be easily applied to excavated soils because methods are available to provide the vigorous mixing needed to combine the binder/stabilizer with the contaminated material. Pretreatment of the waste may be necessary to screen and crush large rocks and debris. Mixing can be performed via in-drum, in-plant, or area-mixing processes. In-drum mixing may be preferred for treatment of small volumes of waste or for toxic wastes. In-plant processes utilize rotary drum mixers for batch processes or pug mill mixers for continuous treatment. Larger volumes of waste may be excavated and moved to a contained area for area mixing. This process involves layering the contaminated material with the stabilizer/binder, and subsequent mixing with a backhoe or similar equipment. Mobile and fixed treatment plants are available for ex situ S/S treatment. Smaller pilot-scale plants can treat up to 100 tons of contaminated soil per day while larger portable plants typically process 500 to over 1000 tons per day [39]. Stabilization/stabilization techniques are available to provide mixing of the binder/stabilizer with the contaminated soil in situ. In situ S/S is less labor and energy intensive than ex situ process that require excavation, transport, and disposal of the treated material. In situ S/S is also preferred if volatile or semivolatile organics are present because excavation would expose these contaminants to the air [85]. However, the presence of bedrock, large boulders cohesive soils, oily sands, and clays may preclude the application of in situ S/S at some sites. It is also more difficult to provide uniform and complete mixing through in situ processes. Mixing of the binder and contaminated matrix may be achieved using in-place mixing, vertical auger mixing, or injection grouting. In-place mixing is similar to ex situ area mixing except that the soil is not excavated prior to treatment. The in situ process is useful for treating surface or shallow contamination and involves spreading and mixing the binders with the waste using conventional excavation equipment such as draglines, backhoes, or clamshell buckets. Vertical auger mixing uses a system of augers to inject and mix the binding reagents with the waste. Larger (6–12 ft diameter) augers are used for shallow (10–40 ft) drilling and can treat 500–1000 cubic yards per day [86, 87]. Deep stabilization/ solidification (up to 150 ft) can be achieved by using ganged augers (up to 3 ft in diameter each) that can treat 150–400 cubic yards per day. Finally injection grouting may be performed to inject the binder containing suspended or dissolved reagents into the treatment area under pressure. The binder permeates the surrounding soil and cures in place [39].

Vitrification

The mobility of metal contaminants can be decreased by high-temperature treatment of the contaminated area that results in the formation of vitreous material, usually an oxide solid. During this process, the increased temperature may also volatilize and/or destroy organic contaminants or volatile metal species (such as Hg) that must be collected for treatment or disposal. Most soils can be treated by vitrification, and a wide variety of inorganic and organic contaminants can be targeted. Vitrification may be performed ex situ or in situ althoughin situ processes are preferred due to the lower energy requirements and cost [88]. Typical stages in ex situvitrification processes may include excavation, pretreatment, mixing, feeding, melting and vitrification, off-gas collection and treatment, and forming or casting of the melted product. The energy requirement for melting is the primary factor influencing the cost of ex situ vitrification. Different sources of energy can be used for this purpose, depending on local energy costs. Process heat losses and water content of the feed should be controlled in order to minimize energy requirements. Vitrified material with certain characteristics may be obtained by using additives such as sand, clay, and/or native soil. The vitrified waste may be recycled and used as clean fill, aggregate, or other reusable materials [39]. In situ vitrification (ISV) involves passing electric current through the soil using an array of electrodes inserted vertically into the contaminated region. Each setting of four electrodes is referred to as a melt. If the soil is too dry, it may not provide sufficient conductance, and a trench containing flaked graphite and glass frit (ground glass particles) must be placed between the electrodes to provide an initial flow path for the current. Resistance heating in the starter path melts the soil. The melt grows outward and down as the molten soil usually provides additional conductance for the current. A single melt can treat up to 1000 tons of contaminated soil to depths of 20 feet, at a typical treatment rate of 3 to 6 tons per hour. Larger areas are treated by fusing together multiple individual vitrification zones. The main requirement for in situ vitrification is the ability of the soil melt to carry current and solidify as it cools. If the alkali content (as Na_2O and K_2O) of the soil is too high (1.4 wt%), the molten soil may not provide enough conductance to carry the current [89].

Vitrification is not a classical immobilization technique. The advantages include (i) easily applied for reclamation of heavily contaminated soils (Pb, Cd, Cr, asbestos, and materials containing asbestos), (ii) in the course of applying this method qualification of wastes (from hazardous to neutral) could be changed.

Assessment of Efficiency and Capacity of Immobilization

The efficiency (E) and capacity (P) of different additives for immobilization and field applications can be evaluated using the expressions

$$E(\%) = \frac{M_o - M_e}{M_o} \times 100,$$

$$P = \frac{(M_o - M_e)V}{m},$$

(2)

where E = efficiency of immobilization agent; P = capacity of immobilization agent; Me = equilibrium extractable concentration of single metal in the immobilized soil (mg L^{-1}); Mo = initial extractable concentration of single metal in preimmobilized soil (mg L^{-1}); V = volume of metal salt solution (mg L^{-1}); m = weight of immobilization agent (g) [90]. High values of E and P represent the perfect efficiency and capacity of an additive that can be used in field studies of metal immobilization. After screening out the best efficient additive, another experiment could be conducted to determine the best ratio (soil/additive) for the field-fixing treatment. After the fixing treatment of contaminated soils, a lot of methods including biological and physiochemical experiments could be used to assess the remediation efficiency. Environmental risk could also be estimated after confirming the immobilized efficiency and possible release [89].

Soil Washing

Soil washing is essentially a volume reduction/waste minimization treatment process. It is done on the excavated (physically removed) soil (ex situ) or on-site (in situ). Soil washing as discussed in this review refers to ex situ techniques that employ physical and/or chemical procedures to extract metal contaminants from soils. During soil washing, (i) those soil particles which host the majority of the contamination are separated from the bulk soil fractions (physical separation), (ii) contaminants are removed from the soil by aqueous chemicals and recovered from solution on a solid substrate (chemical extraction), or (iii) a combination of both [91]. In all cases, the separated contaminants then go to hazardous waste landfill (or occasionally are further treated by chemical, thermal, or biological processes). By removing the majority of the contamination from the soil, the bulk fraction that remains can be (i) recycled on the site being remediated as relatively inert backfill, (ii) used on another site as fill, or (iii) disposed of relatively cheaply as nonhazardous material.

Ex situ soil washing is particularly frequently used in soil remediation because it (i) completely removes the contaminants and hence ensures the rapid cleanup of a contaminated site [92], (ii) meets specific criteria, (iii) reduces or eliminates long-term liability, (iv) may be the most cost-effective solution, and (v) may produce recyclable material or energy [93]. The disadvantages include the fact that the contaminants are simply moved to a different place, where they must be monitored, the risk of spreading contaminated soil and dust particles during removal and transport of contaminated soil, and the relatively high cost. Excavation can be the most expensive option when large amounts of soil must be removed, or disposal as hazardous or toxic waste is required.

Acid and chelator soil washing are the two most prevalent removal methods [94]. Soil washing currently involves soil flushing an in situ process in which the washing solution is forced through the in-place soil matrix, ex situ extraction of heavy metals from the soil slurry in reactors, and soil heap leaching. Another heavy metal removal technology is electroremediation, which mostly involves electrokinetic movement of charged particles suspended in the soil solution, initiated by an electric gradient [35]. The metals can be removed by precipitation at the electrodes. Removal of the majority of the contaminants from the soil does not mean that the contaminant-depleted bulk is totally contaminant free. Thus, for soil washing to be successful, the level of contamination in the treated bulk must be below a site-specific action limit (e.g., based on risk assessment). Cost effectiveness with soil washing is achieved by offsetting processing costs against the ability to significantly reduce the amount of material requiring costly disposal at a hazardous waste landfill [95].

Typically the cleaned fractions from the soil washing process should be >70–80% of the original mass of the soil, but, where the contaminants have a very high associated disposal cost, and/or where transport distances to the nearest hazardous waste landfill are substantial, a 50% reduction might still be cost effective. There is also a generally held opinion that soil washing based on physical separation processes is only cost effective for sandy and granular soils where the clay and silt content (particles less than 0.063 mm) is less than 30–35% of the soil. Soil washing by chemical dissolution of the contaminants is not constrained by the proportion of clay as this fraction can also be leached by the chemical agent. However, clay-rich soils pose other problems such as difficulties with materials handling and solid-liquid separation [96]. Full-scale soil washing plants exist as fixed centralized treatment centres, or as mobile/transportable units. With fixed centralized facilities, contaminated soil is brought to the plant, whereas with mobile/transportable facilities, the plant is transported to a contaminated site, and soil is processed on the site. Where mobile/transportable plant is used, the cost of mobilization and demobilization can be significant. However, where large volumes of soil are to be treated, this cost can be more than offset by reusing clean material on the site (therefore avoiding the cost of transport to an off-site centralized treatment facility, and avoiding the cost of importing clean fill).

Principles of Soil Washing

Soil washing is a volume reduction/waste minimization treatment technology based on physical and/or chemical processes. With physical soil washing, differences between particle grain size, settling velocity, specific gravity, surface chemical behaviour, and rarely magnetic properties are used to separate those particles which host the majority of the contamination from the bulk which are contaminant-depleted. The equipment used is standard mineral processing equipment, which is more generally used in the mining industry [91]. Mineral processing techniques as applied to soil remediation have been reviewed in literature [97].

With chemical soil washing, soil particles are cleaned by selectively transferring the contaminants on the soil into solution. Since heavy metals are sparingly soluble and occur predominantly in a sorbed state, washing the soils with water alone would be expected to remove too low an amount of cations in the leachates, chemical agents have to be added to the washing water [98]. This is achieved by mixing the soil with aqueous solutions of acids, alkalis, complexants, other solvents, and surfactants. The resulting cleaned particles are then separated from the resulting aqueous solution. This solution is then treated to remove the contaminants (e.g., by sorption on activated carbon or ion exchange) [91, 95].

The effectiveness of washing is closely related to the ability of the extracting solution to dissolve the metal contaminants in soils. However, the strong bonds between the soil and metals make the cleaning process difficult [99]. Therefore, only extractants capable of dissolving large quantities of metals would be suitable for cleaning purposes. The realization that the goal of soil remediation is to remove the metal and preserve the natural soil properties limits the choice of extractants that can be used in the cleaning process [100].

Chemical Extractants for Soil Washing

Owing to the different nature of heavy metals, extracting solutions that can optimally remove them must be carefully sought during soil washing. Several classes of chemicals used for soil washing include surfactants, cosolvents, cyclodextrins, chelating agents, and organic acids [101–106]. All these soil washing extractants have been developed on a case-by-case basis depending on the contaminant type at a particular site. A few studies have indicated that the solubilization/exchange/extraction of heavy metals by washing solutions differs considerably for different soil types. Strong acids attack and degrade the soil crystalline structure at extended contact times. For less damaging washes, organic acids and chelating agents are often suggested as alternatives to straight mineral acid use [107].

Natural, low-molecular-weight organic acids (LMWOAs) including oxalic, citric, formic, acetic, malic, succinic, malonic, maleic, lactic, aconitic, and fumaric acids are natural products of root exudates, microbial secretions, and plant and animal residue decomposition in soils [108]. Thus metal dissolution by organic acids is likely to be more representative of a mobile metal fraction that is available to biota [109]. The chelating organic acids are able to dislodge the exchangeable, carbonate, and reducible fractions of heavy metals by washing procedures [94]. Although many chelating compounds including citric acid [108], tartaric acid [110], and EDTA [94, 100, 111] for mobilizing heavy metals have been evaluated, there remain uncertainties as to the optimal choice for full-scale application. The identification and quantification of coexisting solid metal species in the soil before and after treatment are essential to design and assess the efficiency of soil-washing technology [4]. A recent study [112] showed that changes in Ni, Cu, Zn, Cd, and Pb speciation and uptake by maize in a sandy loam before and after washing with three chelating organic acids indicated that EDTA and citric acid appeared to offer greater potentials as chelating agents for remediating the permeable soil. Tartaric acid was, however, recommended in events of moderate contamination.

The use of soil washing to remediate contaminated fine-grained soils that contained more than 30% fines fraction has been reported by several workers [113–115]. Khodadoust et al. [59, 116] have also studied the removal of various metals (Pb, Ni, and Zn) from field and clay (kaolin) soil samples using a broad spectrum of extractants (chelating agents and organic acids). Chen and Hong [117] reported on the chelating extraction of Pb and Cu from an authentic contaminated soil using derivatives of iminodiacetic acid and L-cyestein. Wuana et al. [118] investigated the removal of Pb and Cu from kaolin and bulk clay soils using two mineral acids (HCl and H_2SO_4) and chelating agents (EDTA and oxalic acid). The use of chelating organic acids—citric acid, tartaric acid and EDTA in the simultaneous removal of Ni, Cu, Zn, Cd, and Pb from an experimentally contaminated sandy loam was carried out by Wuana et al. [112]. These studies furnished valuable information on the distribution of heavy metals in the soils and their removal using various extracting solutions.

Phytoremediation

Phytoremediation, also called green remediation, botanoremediation, agroremediation, or vegetative remediation, can be defined as an in situ remediation strategy that uses vegetation and associated microbiota, soil amendments, and agronomic techniques to remove, contain, or render environmental contaminants harmless [119, 120]. The idea of using metal-accumulating plants to remove heavy metals and other compounds was first introduced in 1983, but the concept has actually been implemented for the past 300 years on wastewater discharges [121, 122]. Plants may break down or degrade organic pollutants or remove and stabilize metal contaminants. The methods used to phytoremediate metal contaminants are slightly different from those used to remediate sites polluted with organic contaminants. As it is a relatively new technology, phytoremediation is still mostly in its testing stages and as such has not been used in many places as a full-scale application. However, it has been tested successfully in many places around the world for many different contaminants. Phytoremediation is energy efficient, aesthetically pleasing method of remediating sites with low-to-moderate levels of contamination, and it can be used in conjunction with other more traditional remedial methods as a finishing step to the remedial process.

The advantages of phytoremediation compared with classical remediation are that (i) it is more economically viable using the same tools and supplies as agriculture, (ii) it is less disruptive to the environment and does not involve waiting for new plant communities to recolonize the site, (iii) disposal sites are not needed, (iv) it is more likely to be accepted by the public as it is more aesthetically pleasing then traditional methods, (v) it avoids excavation and transport of polluted media thus reducing the risk of spreading the contamination, and (vi) it has the potential to treat sites polluted with more than one type of pollutant. The disadvantages are as follow (i) it is dependant on the growing conditions required by the plant (i.e., climate, geology, altitude, and temperature), (ii) large-scale operations require access to agricultural equipment and knowledge, (iii) success is dependant on the tolerance of the plant to the pollutant, (iv) contaminants collected in senescing tissues may be released back into the environment in autumn, (v) contaminants may be collected in woody tissues used as fuel, (vi) time taken to remediate sites far exceeds that of other technologies,

(vii) contaminant solubility may be increased leading to greater environmental damage and the possibility of leaching. Potentially useful phytoremediation technologies for remediation of heavy metal-contaminated soils include phytoextraction (phytoaccumulation), phytostabilization, and phytofiltration [123].

Phytoextraction (Phytoaccumulation)

Phytoextraction is the name given to the process where plant roots uptake metal contaminants from the soil and translocate them to their above soil tissues. A plant used for phytoremediation needs to be heavy-metal tolerant, grow rapidly with a high biomass yield per hectare, have high metal-accumulating ability in the foliar parts, have a profuse root system, and a high bioaccumulation factor [21, 124]. Phytoextraction is, no doubt, a publicly appealing (green) remediation technology [125]. Two approaches have been proposed for phytoextraction of heavy metals, namely, continuous or natural phytoextraction and chemically enhanced phytoextraction [126, 127].

Continuous or Natural Phytoextraction

Continuous phytoextraction is based on the use of natural hyperaccumulator plants with exceptional metal-accumulating capacity. Hyperaccumulators are species capable of accumulating metals at levels 100-fold greater than those typically measured in shoots of the common nonaccumulator plants. Thus, a hyperaccumulator plant will concentrate more than $10\,mg\,kg^{-1}$ Hg, $100\,mg\,kg^{-1}$ Cd, $1000\,mg\,kg^{-1}$ Co, Cr, Cu, and Pb; $10\,000\,mg\,kg^{-1}$ Zn and Ni [128, 129]. Hyperaccumulator plant species are used on metalliferous sites due to their tolerance of relatively high levels of pollution. Approximately 400 plant species from at least 45 plant families have been so far, reported to hyperaccumulate metals [22, 127]; some of the families areBrassicaceae, Fabaceae, Euphorbiaceae, Asterraceae, Lamiaceae, and Scrophulariaceae [130, 131]. Crops like alpine pennycress (Thlaspi caerulescens), Ipomea alpine, Haumaniastrum robertii, Astragalus racemosus, Sebertia acuminate have very high bioaccumulation potential for Cd/Zn, Cu, Co, Se, and Ni, respectively [22]. Willow (Salix viminalis L.), Indian mustard (Brassica juncea L.), corn (Zea mays L.), and sunflower (Helianthus annuus L.) have reportedly shown high uptake and tolerance to heavy metals [132]. A list of some plant hyperaccumulators are given in Table 6. A number of processes are involved during phytoextraction of metals from soil: (i) a metal fraction is sorbed at root surface, (ii) bioavailable metal moves across cellular membrane into root cells, (iii) a fraction of the metal absorbed into roots is immobilized in the vacuole, (iv) intracellular mobile metal crosses cellular membranes into root vascular tissue (xylem), and (v) metal is translocated from the root to aerial tissues (stems and leaves) [22]. Once inside the plant, most metals are too insoluble to move freely in the vascular system so they usually form carbonate, sulphate, or phosphate precipitate immobilizing them in apoplastic (extracellular) and symplastic (intracellular) compartments [46]. Hyperaccumulators have several beneficial characteristics but may tend to be slow growing and produce low biomass, and years or decades are needed to clean up contaminated sites. To overcome these shortfalls, chemically enhanced phytoextraction has been developed. The approach makes use of high biomass crops that are induced to take up large amounts of metals when their mobility in soil is enhanced by chemical treatment with chelating organic acids [133].

Table 6: Some metal hyperaccumulating plants [21]

Plant	Metal	Concentration (mg kg−1)
Dicotyledons		
Cystus ladanifer	Cd	309
	Co	2 667
	Cr	2 667
	Ni	4 164
	Zn	7 695
Thlaspi caerulescens	Cd	10 000–15 000
	Zn	10 000–15 000
Arabidopsis halleri	Cd	5 900–31 000
Alyssum sp.	Ni	4 200–24 400
Brassica junica	Pb	10 000–15 000
	Zn	2 600
Betula	Zn	528
Grasses		

Vetiveria zizaniodes	Zn	0.03
Paspalum notatum		
Stenotaphrum secundatum		
Pennisetum glaucum		

Chelate-Assisted (Induced) Phytoextraction

For more than 10 years, chelant-enhanced phytoextraction of metals from contaminated soils have received much attention as a cost-effective alternative to conventional techniques of enhanced soil remediation [133,134]. When the chelating agent is applied to the soil, metal-chelant complexes are formed and taken up by the plant, mostly through a passive apoplastic pathway [133]. Unless the metal ion is transported as a noncationic chelate, apoplastic transport is further limited by the high cation exchange capacity of cell walls [46]. Chelators have been isolated from plants that are strongly involved in the uptake of heavy metals and their detoxification. The chelating agent EDTA has become one of the most tested mobilizing amendments for less mobile/available metals such as Pb [135, 136]. Chelators have been isolated from plants that are strongly involved in the uptake of heavy metals and their detoxification. The addition of EDTA to a Pb-contaminated soil (total soil Pb 2500 mg kg^{-1}) increased shoot lead concentration of Zea mays L. (corn) and Pisun sativum(pea) from less than 500 mg kg^{-1} to more than 10,000 mg kg^{-1}. Enhanced accumulation of metals by plant species with EDTA treatment is attributed to many factors working either singly or in combination. These factors include (i) an increase in the concentration of available metals, (ii) enhanced metal-EDTA complex movement to roots, (iii) less binding of metal-EDTA complexes with the negatively charged cell wall constituents, (iv) damage to physiological barriers in roots either due to greater concentration of metals or EDTA or metal-EDTA complexes, and (v) increased mobility of metals within the plant body when complexed with EDTA compared to free-metal ions facilitating the translocation of metals from roots to shoots [134, 137]. For the chelates tested, the order of effectiveness in increasing Pb desorption from the soil was EDTA > hydroxyethylethylene-diaminetriacetic acid (HEDTA) > diethylenetriaminepentaacetic acid (DTPA) > ethylenediamine di(o-hyroxyphenylacetic acid) EDDHA [135]. Vassil et al. [138] reported that Brassica junceaexposed to Pb and EDTA in hydroponic solution was able to accumulate up to 55 mM kg^{-1} Pb in dry shoot tissue (1.1% w/w). This represents a 75-fold concentration of lead in shoot over that in solution. A 0.25 mM threshold concentration of EDTA was required to stimulate this dramatic accumulation of both lead and EDTA in shoots. Since EDTA has been associated with high toxicity and persistence in the environment, several other alternatives have been proposed. Of all those, EDDS ([S,S]-ethylenediamine disuccinate) has been introduced as a promising and environmentally friendlier mobilizing agent, especially for Cu and Zn [135, 139,140]. Once the plants have grown and absorbed the metal pollutants, they are harvested and disposed of safely. This process is repeated several times to reduce contamination to acceptable levels.

Interestingly, in the last few years, the possibility of planting metal hyperaccumulator crops over a low-grade ore body or mineralized soil, and then harvesting and incinerating the biomass to produce a commercial bio-ore has been proposed [141] though this is usually reserved for use with precious metals. This process calledphytomining offers the possibility of exploiting ore bodies that are otherwise uneconomic to mine, and its effect on the environment is minimal when compared with erosion caused by opencast mining [123, 141].

Assessing the Efficiency of Phytoextraction

Depending on heavy metal concentration in the contaminated soil and the target values sought for in the remediated soil, phytoextraction may involve repeated cropping of the plant until the metal concentration drops to acceptable levels. The ability of the plant to account for the decrease in soil metal concentrations as a function of metal uptake and biomass production plays an important role in achieving regulatory acceptance. Theoretically, metal removal can be accounted for by determining metal concentration in the plant, multiplied by the reduction in soil metal concentrations [127]. It should, however, be borne in mind that this approach may be challenged by a number of factors working together during field applications. Practically, the bioaccumulation factor, f, amount of metal extracted, M (mg/kg plant) and phytoremediation time, tp (year) [142] can be used to evaluate the plant's phytoextraction efficiency and calculated according to equation (3) [143] by assuming that the plant can be cropped n times each year and metal pollution occurs only in the active rooting zone, that is, top soil layer (0–20 cm) and still assuming a soil bulk density of 1.3 t/m^3, giving a total soil mass of 2600 t/ha.

$$f = \frac{\text{Metal concentration in plant shoot}}{\text{Metal concentration in soil}},$$

$$M\,(\text{mg/kg plant}) = \text{Metal concentration in plant tissue} \times \text{Biomass},$$

$$t_p\,(\text{year}) = \frac{\text{Metal concentration in soil needed to decrease} \times \text{Soil mass}}{\text{Metal concentration in plant shoot} \times \text{Plant shoot biomass} \times n}.$$

(3)

Prospects of Phytoextraction

One of the key aspects of the acceptance of phytoextraction pertains to its performance, ultimate utilization of byproducts, and its overall economic viability. Commercialization of phytoextraction has been challenged by the expectation that site remediation should be achieved in a time comparable to other clean-up technologies [123]. Genetic engineering has a great role to play in supplementing the list of plants available for phytoremediation by the use of engineering tools to insert into plants those genes that will enable the plant to metabolize a particular pollutant [144]. A major goal of plant genetic engineering is to enhance the ability of plants to metabolize many of the compounds that are of environmental concern. Currently, some laboratories are using traditional breeding techniques, others are creating protoplast-fusion hybrids, and still others are looking at the direct insertion of novel genes to enhance the metabolic capabilities of plants [144]. On the whole, phytoextraction appears a very promising technology for the removal of metal pollutants from the environment and is at present approaching commercialization.

Possible Utilization of Biomass after Phytoextraction

A serious challenge for the commercialization of phytoextraction has been the disposal of contaminated plant biomass especially in the case of repeated cropping where large tonnages of biomass may be produced. The biomass has to be stored, disposed of or utilized in an appropriate manner so as not to pose any environmental risk. The major constituents of biomass material are lignin, hemicellulose, cellulose, minerals, and ash. It possesses high moisture and volatile matter, low bulk density, and calorific value [127]. Biomass is solar energy fixed in plants in form of carbon, hydrogen, and oxygen (oxygenated hydrocarbons) with a possible general chemical formula $CH_{1.44}O_{0.66}$. Controlled combustion and gasification of biomass can yield a mixture of producer gas and/or pyro-gas which leads to the generation of thermal and electrical energy [145]. Composting and compacting can be employed as volume reduction approaches to biomass reuse [146]. Ashing of biomass can produce bio-ores especially after the phytomining of precious metals. Heavy metals such as Co, Cu, Fe, Mn, Mo, Ni, and Zn are plant essential metals, and most plants have the ability to accumulate them [147]. The high concentrations of these metals in the harvested biomass can be "diluted" to acceptable concentrations by combining the biomass with clean biomass in formulations of fertilizer and fodder.

Phytostabilization

Phytostabilization, also referred to as in-place inactivation, is primarily concerned with the use of certain plants to immobilize soil sediment and sludges [148]. Contaminant are absorbed and accumulated by roots, adsorbed onto the roots, or precipitated in the rhizosphere. This reduces or even prevents the mobility of the contaminants preventing migration into the groundwater or air and also reduces the bioavailability of the contaminant thus preventing spread through the food chain. Plants for use in phytostabilization should be able to (i) decrease the amount of water percolating through the soil matrix, which may result in the formation of a hazardous leachate, (ii) act as barrier to prevent direct contact with the contaminated soil, and (iii) prevent soil erosion and the distribution of the toxic metal to other areas [46]. Phytostabilization can occur through the process of sorption, precipitation, complexation, or metal valence reduction. This technique is useful for the cleanup of Pb, As, Cd, Cr, Cu, and Zn [147]. It can also be used to reestablish a plant community on sites that have been denuded due to the high levels of metal contamination. Once a community of tolerant species has been established, the potential for wind erosion (and thus spread of the pollutant) is reduced, and leaching of the soil contaminants is also reduced. Phytostabilization is advantageous because disposal of hazardous material/biomass is not required, and it is very effective when rapid immobilization is needed to preserve ground and surface waters [147, 148].

Phytofiltration

Phytofiltration is the use of plant roots (rhizofiltration) or seedlings (blastofiltration), is similar in concept to phytoextraction, but is used to absorb or adsorb pollutants, mainly metals, from groundwater and aqueous-waste streams rather than the remediation of polluted soils [3, 123]. Rhizosphere is the soil area immediately surrounding the plant root surface, typically up to a few millimetres from the root surface. The contaminants are either adsorbed onto the root surface or are absorbed by the plant roots. Plants used for rhizofiltration are not planted directly in situ but are acclimated to the pollutant first. Plants are hydroponically grown in clean water rather than soil, until a large root system has developed. Once a large root system is in place, the water supply is substituted for a polluted water supply to acclimatize the plant. After the plants become acclimatized, they are planted in the polluted area where the roots uptake the polluted water and the contaminants along with it. As the roots become saturated, they are harvested and disposed of safely. Repeated treatments of the site can reduce pollution to suitable levels as was exemplified in Chernobyl where sunflowers were grown in radioactively contaminated pools [21].

CONCLUSIONS

Background knowledge of the sources, chemistry, and potential risks of toxic heavy metals in contaminated soils is necessary for the selection of appropriate remedial options. Remediation of soil contaminated by heavy metals is necessary in order to reduce the associated risks, make the land resource available for agricultural production, enhance food security, and scale down land tenure problems. Immobilization, soil washing, and phytoremediation are frequently listed among the best available technologies for cleaning up heavy metal contaminated soils but have been mostly demonstrated in developed countries. These technologies are recommended for field applicability and commercialization in the developing countries also where agriculture, urbanization, and industrialization are leaving a legacy of environmental degradation.

REFERENCES

1. S. Khan, Q. Cao, Y. M. Zheng, Y. Z. Huang, and Y. G. Zhu, "Health risks of heavy metals in contaminated soils and food crops irrigated with wastewater in Beijing, China," Environmental Pollution, vol. 152, no. 3, pp. 686–692, 2008. · · ·

2. M. K. Zhang, Z. Y. Liu, and H. Wang, "Use of single extraction methods to predict bioavailability of heavy metals in polluted soils to rice," Communications in Soil Science and Plant Analysis, vol. 41, no. 7, pp. 820–831, 2010. · ·

3. GWRTAC, "Remediation of metals-contaminated soils and groundwater," Tech. Rep. TE-97-01,, GWRTAC, Pittsburgh, Pa, USA, 1997, GWRTAC-E Series.

4. T. A. Kirpichtchikova, A. Manceau, L. Spadini, F. Panfili, M. A. Marcus, and T. Jacquet, "Speciation and solubility of heavy metals in contaminated soil using X-ray microfluorescence, EXAFS spectroscopy, chemical extraction, and thermodynamic modeling," Geochimica et Cosmochimica Acta, vol. 70, no. 9, pp. 2163–2190, 2006. · ·

5. D. C. Adriano, Trace Elements in Terrestrial Environments: Biogeochemistry, Bioavailability and Risks of Metals, Springer, New York, NY, USA, 2nd edition, 2003.

6. P. Maslin and R. M. Maier, "Rhamnolipid-enhanced mineralization of phenanthrene in organic-metal co-contaminated soils," Bioremediation Journal, vol. 4, no. 4, pp. 295–308, 2000.

7. M. J. McLaughlin, B. A. Zarcinas, D. P. Stevens, and N. Cook, "Soil testing for heavy metals," Communications in Soil Science and Plant Analysis, vol. 31, no. 11–14, pp. 1661–1700, 2000.

8. M. J. McLaughlin, R. E. Hamon, R. G. McLaren, T. W. Speir, and S. L. Rogers, "Review: a bioavailability-based rationale for controlling metal and metalloid contamination of agricultural land in Australia and New Zealand," Australian Journal of Soil Research, vol. 38, no. 6, pp. 1037–1086, 2000. · ·

9. W. Ling, Q. Shen, Y. Gao, X. Gu, and Z. Yang, "Use of bentonite to control the release of copper from contaminated soils," Australian Journal of Soil Research, vol. 45, no. 8, pp. 618–623, 2007. ··

10. A. Kabata-Pendias and H. Pendias, Trace Metals in Soils and Plants, CRC Press, Boca Raton, Fla, USA, 2nd edition, 2001.

11. Q. Zhao and J. J. Kaluarachchi, "Risk assessment at hazardous waste-contaminated sites with variability of population characteristics," Environment International, vol. 28, no. 1-2, pp. 41–53, 2002. ··

12. N. S. Bolan, B.G. Ko, C.W.N. Anderson, and I. Vogeler, "Solute interactions in soils in relation to bioavailability and remediation of the environment," in Proceedings of the 5th International Symposium of Interactions of Soil Minerals with Organic Components and Microorganisms, Pucón, Chile, November 2008.

13. G. M. Pierzynski, J. T. Sims, and G. F. Vance, Soils and Environmental Quality, CRC Press, London, UK, 2nd edition, 2000.

14. J. J. D›Amore, S. R. Al-Abed, K. G. Scheckel, and J. A. Ryan, "Methods for speciation of metals in soils: a review," Journal of Environmental Quality, vol. 34, no. 5, pp. 1707–1745, 2005. ···

15. B. J. Alloway, Heavy Metals in Soils, Blackie Academic and Professional, London, UK, 2nd edition, 1995.

16. E. Lombi and M. H. Gerzabek, "Determination of mobile heavy metal fraction in soil: results of a pot experiment with sewage sludge," Communications in Soil Science and Plant Analysis, vol. 29, no. 17-18, pp. 2545–2556, 1998.

17. G. Sposito and A. L. Page, "Cycling of metal ions in the soil environment," in Metal Ions in Biological Systems, H. Sigel, Ed., vol. 18 of Circulation of Metals in the Environment, pp. 287–332, Marcel Dekker, Inc., New York, NY, USA, 1984.

18. S. Kuo, P. E. Heilman, and A. S. Baker, "Distribution and forms of copper, zinc, cadmium, iron, and manganese in soils near a copper smelter," Soil Science, vol. 135, no. 2, pp. 101–109, 1983.

19. M. Kaasalainen and M. Yli-Halla, "Use of sequential extraction to assess metal partitioning in soils,"Environmental Pollution, vol. 126, no. 2, pp. 225–233, 2003. ·

20. N. T. Basta, J. A. Ryan, and R. L. Chaney, "Trace element chemistry in residual-treated soil: key concepts and metal bioavailability," Journal of Environmental Quality, vol. 34, no. 1, pp. 49–63, 2005.

21. A. Scragg, Environmental Biotechnology, Oxford University Press, Oxford, UK, 2nd edition, 2006.

22. M.M. Lasat, "Phytoextraction of metals from contaminated soil: a review of plant/soil/metal interaction and assessment of pertinent agronomic issues," Journal of Hazardous Substances Research, vol. 2, pp. 1–25, 2000.

23. L. H. P. Jones and S. C. Jarvis, "The fate of heavy metals," in The Chemistry of Soil Processes, D. J. Green and M. H. B. Hayes, Eds., p. 593, John Wiley & Sons, New York, NY, USA, 1981.

24. P. H. Raven, L. R. Berg, and G. B. Johnson, Environment, Saunders College Publishing, New York, NY, USA, 2nd edition, 1998.

25. M. E. Sumner, "Beneficial use of effluents, wastes, and biosolids," Communications in Soil Science and Plant Analysis, vol. 31, no. 11–14, pp. 1701–1715, 2000.

26. R. L. Chaney and D. P. Oliver, "Sources, potential adverse effects and remediation of agricultural soil contaminants," in Contaminants and the Soil Environments in the Australia-Pacific Region, R. Naidu, Ed., pp. 323–359, Kluwer Academic Publishers, Dordrecht, The Netherlands, 1996.

27. USEPA, "A plain english guide to the EPA part 503 biosolids rule," USEPA Rep. 832/R-93/003, USEPA, Washington, DC, USA, 1994.

28. K. Weggler, M. J. McLaughlin, and R. D. Graham, "Effect of Chloride in Soil Solution on the Plant Availability of Biosolid-Borne Cadmium," Journal of Environmental Quality, vol. 33, no. 2, pp. 496–504, 2004.

29. M. L. A. Silveira, L. R. F. Alleoni, and , and L. R. G. Guilherme, "Biosolids and heavy metals in soils,"Scientia Agricola, vol. 60, no. 4, pp. 64–111, 2003.

30. R. Canet, F. Pomares, F. Tarazona, and M. Estela, "Sequential fractionation and plant availability of heavy metals as affected by sewage sludge applications to soil," Communications in Soil Science and Plant Analysis, vol. 29, no. 5-6, pp. 697–716, 1998.

31. S. V. Mattigod and A. L. Page, "Assessment of metal pollution in soil," in Applied Environmental Geochemistry, pp. 355–394, Academic Press, London, UK, 1983.

32. R. G. McLaren, L. M. Clucas, and M. D. Taylor, "Leaching of macronutrients and metals from undisturbed soils treated with metal-spiked sewage sludge. 3. Distribution of residual metals,"Australian Journal of Soil Research, vol. 43, no. 2, pp. 159–170, 2005. ··

33. C. Keller, S. P. McGrath, and S. J. Dunham, "Trace metal leaching through a soil-grassland system after sewage sludge application," Journal of Environmental Quality, vol. 31, no. 5, pp. 1550–1560, 2002.

34. R. G. McLaren, L. M. Clucas, M. D. Taylor, and T. Hendry, "Leaching of macronutrients and metals from undisturbed soils treated with metal-spiked sewage sludge. 2. Leaching of metals," Australian Journal of Soil Research, vol. 42, no. 4, pp. 459–471, 2004. ··

35. S. C. Reed, R. W. Crites, and E. J. Middlebrooks, Natural Systems for Waste Management and Treatment, McGraw-Hill, New York, NY, USA, 2nd edition, 1995.

36. J. Bjuhr, Trace Metals in Soils Irrigated with Waste Water in a Periurban Area Downstream Hanoi City, Vietnam, Seminar Paper, Institutionen för markvetenskap, Sveriges lantbruksuniversitet (SLU), Uppsala, Sweden, 2007.

37. P. S. DeVolder, S. L. Brown, D. Hesterberg, and K. Pandya, "Metal bioavailability and speciation in a wetland tailings repository amended with biosolids compost, wood ash, and sulfate," Journal of Environmental Quality, vol. 32, no. 3, pp. 851–864, 2003.

38. N. T. Basta and R. Gradwohl, "Remediation of heavy metal-contaminated soil using rock phosphate,"Better Crops, vol. 82, no. 4, pp. 29–31, 1998.

39. L. A. Smith, J. L. Means, A. Chen, et al., Remedial Options for Metals-Contaminated Sites, Lewis Publishers, Boca Raton, Fla, USA,, 1995.

40. USEPA, Report: recent Developments for In Situ Treatment of Metals contaminated Soils, U.S. Environmental Protection Agency, Office of Solid Waste and Emergency Response, 1996.

41. J. Shiowatana, R. G. McLaren, N. Chanmekha, and A. Samphao, "Fractionation of arsenic in soil by a continuous-flow sequential extraction method," Journal of Environmental Quality, vol. 30, no. 6, pp. 1940–1949, 2001.

42. J. Buekers, Fixation of cadmium, copper, nickel and zinc in soil: kinetics, mechanisms and its effect on metal bioavailability, Ph.D. thesis, Katholieke Universiteit Lueven, 2007, Dissertationes De Agricultura, Doctoraatsprooefschrift nr.

43. D. B. Levy, K. A. Barbarick, E. G. Siemer, and L. E. Sommers, "Distribution and partitioning of trace metals in contaminated soils near Leadville, Colorado," Journal of Environmental Quality, vol. 21, no. 2, pp. 185–195, 1992.

44. USDHHS, Toxicological profile for lead, United States Department of Health and Human Services, Atlanta, Ga, USA, 1999.

45. S.E. Manahan, Toxicological Chemistry and Biochemistry, CRC Press, Limited Liability Company (LLC), 3rd edition, 2003.

46. I. Raskin and B. D. Ensley, Phytoremediation of Toxic Metals: Using Plants to Clean Up the Environment, John Wiley & Sons, New York, NY, USA, 2000.

47. NSC, Lead Poisoning, National Safety Council, 2009, http://www. nsc.org/news_resources/Resources/Documents/Lead_Poisoning. pdf.

48. D. R. Baldwin and W. J. Marshall, "Heavy metal poisoning and its laboratory investigation," Annals of Clinical Biochemistry, vol. 36, no. 3, pp. 267–300, 1999.

49. C.J. Rosen, Lead in the home garden and urban soil environment, Communication and Educational Technology Services, University of Minnesota Extension, 2002.

50. P. Chrostowski, J. L. Durda, and K. G. Edelmann, "The use of natural processes for the control of chromium migration," Remediation, vol. 2, no. 3, pp. 341–351, 1991.

51. I. Bodek, W. J. Lyman, W. F. Reehl, and D. H. Rosenblatt, in Environmental Inorganic Chemistry: Properties, Processes and Estimation Methods, Pergamon Press, Elmsford, NY, USA, 1988.

52. B. E. Davies and L. H. P. Jones, "Micronutrients and toxic elements," in Russell›s Soil Conditions and Plant Growth, A. Wild, Ed., pp. 781–814, John Wiley & Sons; Interscience, New York, NY, USA, 11th edition, 1988.

53. K. M. Greany, An assessment of heavy metal contamination in the marine sediments of Las Perlas Archipelago, Gulf of Panama, M.S. thesis, School of Life Sciences Heriot-Watt University, Edinburgh, Scotland, 2005.

54. P. G. C. Campbell, "Cadmium-A priority pollutant," Environmental Chemistry, vol. 3, no. 6, pp. 387–388, 2006. ··

55. VCI, Copper history/Future, Van Commodities Inc., 2011,http://trademetalfutures.com/copperhistory.html.

56. C. E. Martínez and H. L. Motto, "Solubility of lead, zinc and copper added to mineral soils,"Environmental Pollution, vol. 107, no. 1, pp. 153–158, 2000. ··

57. J. Eriksson, A. Andersson, and R. Andersson, "The state of Swedish farmlands," Tech. Rep. 4778, Swedish Environmental Protection Agency, Stockholm, Sweden, 1997.

58. M. Pourbaix, Atlas of Electrochemical Equilibria, Pergamon Press, New York, NY, USA, 1974, Translated from French by J.A. Franklin.

59. A. P. Khodadoust, K. R. Reddy, and K. Maturi, "Removal of nickel and phenanthrene from kaolin soil using different extractants," Environmental Engineering Science, vol. 21, no. 6, pp. 691–704, 2004. ··

60. DPR-EGASPIN, Environmental Guidelines and Standards for the Petroleum Industry in Nigeria (EGASPIN), Department of Petroleum Resources, Lagos, Nigeria, 2002.

61. A. Tessier, P. G. C. Campbell, and M. Blsson, "Sequential extraction procedure for the speciation of particulate trace metals," Analytical Chemistry, vol. 51, no. 7, pp. 844–851, 1979.

62. A. M. Ure, PH. Quevauviller, H. Muntau, and B. Griepink, "Speciation of heavy metals in soils and sediments. An account of the improvement and harmonization of extraction techniques undertaken under the auspices of the BCR of Commission of the European Communities," International Journal of Environmental Analytical Chemistry, vol. 51, no. 1, pp. 35–151, 1993.

63. D. M. DiToro, J. D. Mahony, D. J. Hansen, K. J. Scott, A. R. Carlson, and G. T. Ankley, "Acid volatile sulfide predicts the acute toxicity of cadmium and nickel in sediments," Environmental Science and Technology, vol. 26, no. 1, pp. 96–101, 1992.

64. R. G. Riley, J. M. Zachara, and F. J. Wobber, "Chemical contaminants on DOE lands and selection of contaminated mixtures for subsurface science research," US-DOE, Energy Resource Subsurface Science Program, Washington, DC, USA, 1992.

65. NJDEP, Soil Cleanup Criteria, New Jersey Department of Environmental Protection, Proposed Cleanup Standards for Contaminated Sites, NJAC 7:26D, 1996.

66. T. A. Martin and M. V. Ruby, "Review of in situ remediation technologies for lead, zinc and cadmium in soil," Remediation, vol. 14, no. 3, pp. 35–53, 2004.

67. S. K. Gupta, T. Herren, K. Wenger, R. Krebs, and T. Hari, "In situ gentle remediation measures for heavy metal-polluted soils," in Phytoremediation of Contaminated Soil and Water, N. Terry and G. Bañuelos, Eds., pp. 303–322, Lewis Publishers, Boca Raton, Fla, USA, 2000.

68. USEPA, "Treatment technologies for site cleanup: annual status report (12th Edition)," Tech. Rep. EPA-542-R-07-012, Solid Waste and Emergency Response (5203P), Washington, DC, USA, 2007.

69. USEPA, "Recent developments for in situ treatment of metal contaminated soils," Tech. Rep. EPA-542-R-97-004, USEPA, Washington, DC, USA, 1997.

70. Y. Hashimoto, H. Matsufuru, M. Takaoka, H. Tanida, and T. Sato, "Impacts of chemical amendment and plant growth on lead speciation and enzyme activities in a shooting range soil: an X-ray absorption fine structure investigation," Journal of Environmental Quality, vol. 38, no. 4, pp. 1420–1428, 2009. $\cdots$

71. N. Finžgar, B. Kos, and D. Leštan, "Bioavailability and mobility of Pb after soil treatment with different remediation methods," Plant, Soil and Environment, vol. 52, no. 1, pp. 25–34, 2006.

72. J. Boisson, M. Mench, J. Vangronsveld, A. Ruttens, P. Kopponen, and T. De Koe, "Immobilization of trace metals and arsenic by different soil additives: evaluation by means of chemical extractions,"Communications in Soil Science and Plant Analysis, vol. 30, no. 3-4, pp. 365–387, 1999.

73. E. Lombi, F. J. Zhao, G. Zhang et al., "In situ fixation of metals in soils using bauxite residue: chemical assessment," Environmental Pollution, vol. 118, no. 3, pp. 435–443, 2002. $\cdots$

74. C. O. Anoduadi, L. B. Okenwa, F. E. Okieimen, A. T. Tyowua, and E.G. Uwumarongie-Ilori, "Metal immobilization in CCA contaminated soil using laterite and termite mound soil. Evaluation by chemical fractionation," Nigerian Journal of Applied Science, vol. 27, pp. 77–87, 2009.

75. L. Q. Wang, L. Luo, Y. B Ma, D. P. Wei, and L. Hua, "In situ immobilization remediation of heavy metals-contaminated soils: a review," Chinese Journal of Applied Ecology, vol. 20, no. 5, pp. 1214–1222, 2009.

76. F. R. Evanko and D. A. Dzombak, "Remediation of metals contaminated soils and groundwater," Tech. Rep. TE-97-01, Groundwater Remediation Technologies Analysis Centre, Pittsburg, Pa, USA, 1997.

77. E. M. Fawzy, "Soil remediation using in situ immobilisation techniques," Chemistry and Ecology, vol. 24, no. 2, pp. 147–156, 2008. · ·

78. M. Farrell, W. T. Perkins, P. J. Hobbs, G. W. Griffith, and D. L. Jones, "Migration of heavy metals in soil as influenced by compost amendments," Environmental Pollution, vol. 158, no. 1, pp. 55–64, 2010. · · ·

79. P. Bishop, D. Gress, and J. Olafsson, "Cement stabilization of heavy metals:Leaching rate assessment," inIndustrial Wastes-Proceedings of the 14th Mid-Atlantic Industrial Waste Conference, Technomics, Lancaster, Pa, USA, 1982.

80. W. Shively, P. Bishop, D. Gress, and T. Brown, "Leaching tests of heavy metals stabilized with Portland cement," Journal of the Water Pollution Control Federation, vol. 58, no. 3, pp. 234–241, 1986.

81. USEPA, "Interference mechanisms in waste stabilization/solidification processes," Tech. Rep. EPA/540/A5-89/004, United States Environmental Protection Agency, Office of Research and Development, Cincinnati, Ohio, USA, 1990.

82. G. Guo, Q. Zhou, and L. Q. Ma, "Availability and assessment of fixing additives for the in situ remediation of heavy metal contaminated soils: a review," Environmental Monitoring and Assessment, vol. 116, no. 1–3, pp. 513–528, 2006. · ·

83. J. R. Conner, Chemical Fixation and Solidification of Hazardous Wastes, Van Nostrand Reinhold, New York, NY, USA, 1990.

84. USEPA, "Stabilization/solidification of CERCLA and RCRA wastes," Tech. Rep. EPA/625/6-89/022, United States Environmental Protection Agency, Center for Environmental Research Information, Cincinnati, Ohio, USA, 1989.

85. USEPA, "International waste technologies/geo-con in situ stabilization/solidification," Tech. Rep. EPA/540/A5-89/004, United States Environmental Protection Agency, Office of Research and Development, Cincinnati, Ohio, USA, 1990.

86. B. H. Jasperse and C. R. Ryan, "Stabilization and fixation using soil mixing," in Proceedings of the ASCE Specialty Conference on Grouting, Soil Improvement, and Geosynthetics, ASCE Publications, Reston, Va, USA, 1992.

87. C. R. Ryan and A. D. Walker, "Soil mixing for soil improvement," in Proceedings of the 23rd Conference on In situ Soil Modification, Geo-Con, Inc., Louisville, Ky, USA, 1992.

88. USEPA, "Vitrification technologies for treatment of Hazardous and radioactive waste handbook," Tech. Rep. EPA/625/R-92/002, United States Environmental Protection Agency, Office of Research and Development, Washington, DC, USA, 1992.

89. J. L. Buelt and L. E. Thompson, The In situ Vitrification Integrated Program: Focusing on an Innovative Solution on Environmental Restoration Needs, Battelle Pacific Northwest Laboratory, Richland, Wash, USA, 1992.

90. A. Jang, Y. S. Choi, and I. S. Kim, "Batch and column tests for the development of an immobilization technology for toxic heavy metals in contaminated soils of closed mines," Water Science and Technology, vol. 37, no. 8, pp. 81–88, 1998. ··

91. G. Dermont, M. Bergeron, G. Mercier, and M. Richer-Laflèche, "Soil washing for metal removal: a review of physical/chemical technologies and field applications," Journal of Hazardous Materials, vol. 152, no. 1, pp. 1–31, 2008. ···

92. P. Wood, "Remediation methods for contaminated sites," in Contaminated Land and Its Reclamation, R. Hester and R. Harrison, Eds., Royal Society of Chemistry, Cambridge, UK, 1997.

93. GOC, "Site Remediation Technologies: A Reference Manual," 2003, Contaminated Sites Working Group, Government of Canada, Ontario, Canada.

94. R. W. Peters, "Chelant extraction of heavy metals from contaminated soils," Journal of Hazardous Materials, vol. 66, no. 1-2, pp. 151–210, 1999. ··

95. CLAIRE, "Understanding soil washing, contaminated land: applications in real environments," Tech. Rep. TB13, 2007.

96. M. Pearl and P. Wood, "Review of pilot and full scale soil washing plants," Warren Spring Laboratory Report LR 1018, Department of the Environment, AEA Technology National Environmental Technology Centre, 1994, B551 Harwell, Oxfordshire, OX11 0RA.

97. A. Gosselin, M. Blackburn, and M. Bergeron, Assessment Protocol of the applicability of ore-processing technology to Treat Contaminated Soils, Sediments and Sludges, prepared for Eco-Technology innovation Section, Eco-Technology Innovation Section, Technology Development and Demonstration Program, Environment Canada, Canada, 1999.

98. A. P. Davis and I. Singh, "Washing of zinc(II) from contaminated soil column," Journal of Environmental Engineering, vol. 121, no. 2, pp. 174–185, 1995. ··

99. D. Gombert, "Soil washing and radioactive contamination," Environmental Progress, vol. 13, no. 2, pp. 138–142, 1994.

100. R. S. Tejowulan and W. H. Hendershot, "Removal of trace metals from contaminated soils using EDTA incorporating resin trapping techniques," Environmental Pollution, vol. 103, no. 1, pp. 135–142, 1998. ··

101. USEPA, "Engineering bulletin: soil washing treatment," Tech. Rep. EPA/540/2-90/017, Office of Emergency and Remedial Response, United States Environmental Protection Agency, Washington, DC, USA, 1990.

102. A. L. Wood, D. C. Bouchard, M. L. Brusseau, and P. S. C. Rao, "Cosolvent effects on sorption and mobility of organic contaminants in soils," Chemosphere, vol. 21, no. 4-5, pp. 575–587, 1990.

103. W. Chu and K. H. Chan, "The mechanism of the surfactant-aided soil washing system for hydrophobic and partial hydrophobic organics," Science of the Total Environment, vol. 307, no. 1–3, pp. 83–92, 2003. ···

104. Y. Gao, J. He, W. Ling, H. Hu, and F. Liu, "Effects of organic acids on copper and cadmium desorption from contaminated soils," Environment International, vol. 29, no. 5, pp. 613–618, 2003. · · ·

105. K. Maturi and K. R. Reddy, "Extractants for the removal of mixed contaminants from soils," Soil and Sediment Contamination, vol. 17, no. 6, pp. 586–608, 2008. · ·

106. H. Zhang, Z. Dang, L. C. Zheng, and X. Y. Yi, "Remediation of soil co-contaminated with pyrene and cadmium by growing maize (Zea mays L.)," International Journal of Environmental Science and Technology, vol. 6, no. 2, pp. 249–258, 2009.

107. J. Yu and D. Klarup, "Extraction kinetics of copper, zinc, iron, and manganese from contaminated sediment using disodium ethylenediaminetetraacetate," Water, Air, and Soil Pollution, vol. 75, no. 3-4, pp. 205–225, 1994.

108. R. Naidu and R. D. Harter, "Effect of different organic ligands on cadmium sorption by and extractability from soils," Soil Science Society of America Journal, vol. 62, no. 3, pp. 644–650, 1998.

109. J. Labanowski, F. Monna, A. Bermond et al., "Kinetic extractions to assess mobilization of Zn, Pb, Cu, and Cd in a metal-contaminated soil: EDTA vs. citrate," Environmental Pollution, vol. 152, no. 3, pp. 693–701, 2008. · · ·

110. X. Ke, P. J. Li, Q. X. Zhou, Y. Zhang, and T. H. Sun, "Removal of heavy metals from a contaminated soil using tartaric acid," Journal of Environmental Sciences, vol. 18, no. 4, pp. 727–733, 2006.

111. B. Sun, F. J. Zhao, E. Lombi, and S. P. McGrath, "Leaching of heavy metals from contaminated soils using EDTA," Environmental Pollution, vol. 113, no. 2, pp. 111–120, 2001. · ·

112. R. A. Wuana, F. E. Okieimen, and J. A. Imborvungu, "Removal of heavy metals from a contaminated soil using organic chelating acids," International Journal of Environmental Science and Technology, vol. 7, no. 3, pp. 485–496, 2010.

113. H. Farrah and W. F. Pickering, "Extraction of heavy metal ions sorbed on clays," Water, Air, and Soil Pollution, vol. 9, no. 4, pp. 491–498, 1978.

114. B. J. W. Tuin and M. Tels, "Removing heavy metals from contaminated clay soils by extraction with hydrochloric acid, edta or hypochlorite solutions," Environmental Technology, vol. 11, no. 11, pp. 1039–1052, 1990.

115. K. R. Reddy and S. Chinthamreddy, "Comparison of extractants for removing heavy metals from contaminated clayey soils," Soil and Sediment Contamination, vol. 9, no. 5, pp. 449–462, 2000.

116. A. P. Khodadoust, K. R. Reddy, and K. Maturi, "Effect of different extraction agents on metal and organic contaminant removal from a field soil," Journal of Hazardous Materials, vol. 117, no. 1, pp. 15–24, 2005. · · ·

117. T. C. Chen and A. Hong, "Chelating extraction of lead and copper from an authentic contaminated soil using N-(2-acetamido) iminodiacetic acid and S-carboxymethyl-L-cysteine," Journal of Hazardous Materials, vol. 41, no. 2-3, pp. 147–160, 1995. · ·

118. R. A. Wuana, F. E. Okieimen, and R. E. Ikyereve, "Removal of lead and copper from contaminated kaolin and bulk clay soils using acids and chelating agents," Journal of Chemical Society of Nigeria, vol. 33, no. 1, pp. 213–219, 2008.

119. S. D. Cunningham and D. W. Ow, "Promises and prospects of phytoremediation," Plant Physiology, vol. 110, no. 3, pp. 715–719, 1996.

120. H. S. Helmisaari, M. Salemaa, J. Derome, O. Kiikkilä, C. Uhlig, and T. M. Nieminen, "Remediation of heavy metal-contaminated forest soil using recycled organic matter and native woody plants," Journal of Environmental Quality, vol. 36, no. 4, pp. 1145–1153, 2007. · ··

121. R. L. Chaney, M. Malik, Y. M. Li et al., "Phytoremediation of soil metals," Current Opinion in Biotechnology, vol. 8, no. 3, pp. 279–284, 1997. · ·

122. R. J. Henry, An Overview of the Phytoremediation of Lead and Mercury, United States Environmental Protection Agency Office of Solid Waste and Emergency Response Technology Innovation office, Washington, DC, USA, 2000.

123. C. Garbisu and I. Alkorta, "Phytoextraction: a cost-effective plant-based technology for the removal of metals from the environment," Bioresource Technology, vol. 77, no. 3, pp. 229–236, 2001. · ·

124. C. D. Jadia and M. H. Fulekar, "Phytotoxicity and remediation of heavy metals by fibrous root grass (sorghum)," Journal of Applied Biosciences, vol. 10, no. 1, pp. 491–499, 2008.

125. M. Vysloužilová, P. Tlustoš, J. Száková, and D. Pavlíková, "As, Cd, Pb and Zn uptake by Salix spp. clones grown in soils enriched by high loads of these elements," Plant, Soil and Environment, vol. 49, no. 5, pp. 191–196, 2003.

126. E. Lombi, F. J. Zhao, S. J. Dunham, and S. P. McGrath, "Phytoremediation of heavy metal-contaminated soils: natural hyperaccumulation versus chemically enhanced phytoextraction," Journal of Environmental Quality, vol. 30, no. 6, pp. 1919–1926, 2001.

127. M. Ghosh and S. P. Singh, "A review on phytoremediation of heavy metals and utilization of its byproducts," Applied Ecology and Environmental Research, vol. 3, no. 1, pp. 1–18, 2005.

128. A. J. M. Baker and R. R. Brooks, "Terrestrial higher plants which hyperaccumulate metallic elements: a review of their distribution, ecology and phytochemistry," Biorecovery, vol. 1, pp. 81–126, 1989.

129. M. M. Lasat, "Phytoextraction of toxic metals: a review of biological mechanisms," Journal of Environmental Quality, vol. 31, no. 1, pp. 109–120, 2002.

130. D. E. Salt, R. D. Smith, and I. Raskin, "Phytoremediation," Annual Reviews in Plant Physiology & Plant Molecular Biology, vol. 49, pp. 643–668, 1998.

131. S. Dushenkov, "Trends in phytoremediation of radionuclides," Plant and Soil, vol. 249, no. 1, pp. 167–175, 2003. ··

132. U. Schmidt, "Enhancing phytoextraction: the effect of chemical soil manipulation on mobility, plant accumulation and leaching of heavy metals," Journal of Environmental Quality, vol. 32, no. 6, pp. 1939–1954, 2003.

133. B. Nowack, R. Schulin, and B. H. Robinson, "Critical assessment of chelant-enhanced metal phytoextraction," Environmental Science and Technology, vol.40, no. 17, pp.5225–5232,2006. ··

134. M. W. H. Evangelou, M. Ebel, and A. Schaeffer, "Chelate assisted phytoextraction of heavy metals from soil. Effect, mechanism, toxicity, and fate of chelating agents," Chemosphere, vol. 68, no. 6, pp. 989–1003, 2007. ···

135. J. W. Huang, J. Chen, W. R. Berti, and S. D. Cunningham, "Phytoremediadon of lead-contaminated soils: role of synthetic chelates in lead phytoextraction," Environmental Science and Technology, vol. 31, no. 3, pp. 800–805, 1997. ··

136. Saifullah, E. Meers, M. Qadir, et al., "EDTA-assisted Pb phytoextraction," Chemosphere, vol. 74, no. 10, pp. 1279–1291, 2009.

137. Y. Xu, N. Yamaji, R. Shen, and J. F. Ma, "Sorghum roots are inefficient in uptake of EDTA-chelated lead," Annals of Botany, vol. 99, no. 5, pp. 869–875, 2007.

138. A. D. Vassil, Y. Kapulnik, I. Raskin, and D. E. Sait, "The role of EDTA in lead transport and accumulation by Indian mustard," Plant Physiology, vol. 117, no. 2, pp. 447–453, 1998.

139. B. Kos and D. Leštan, "Chelator induced phytoextraction and in situ soil washing of Cu," Environmental Pollution, vol. 132, no. 2, pp. 333–339, 2004. ···

140. S. Tandy, K. Bossart, R. Mueller et al., "Extraction of heavy metals from soils using biodegradable chelating agents," Environmental Science and Technology, vol. 38, no. 3, pp. 937–944, 2004. ··

141. R. R. Brooks, M. F. Chambers, L. J. Nicks, and B. H. Robinson, "Phytomining," Trends in Plant Science, vol. 3, no. 9, pp. 359–362, 1998. ··

142. P. Zhuang, Z. H. Ye, C. Y. Lan, Z. W. Xie, and W. S. Shu, "Chemically assisted phytoextraction of heavy metal contaminated soils using three plant species," Plant and Soil, vol. 276, no. 1-2, pp. 153–162, 2005. ··

143. X. Zhang, H. Xia, Z. Li, P. Zhuang, and B. Gao, "Potential of four forage grasses in remediation of Cd and Zn contaminated soils," Bioresource Technology, vol. 101, no. 6, pp. 2063–2066, 2010. ···

144. L. A. Newman, S. E. Strand, N. Choe et al., "Uptake and biotransformation of trichloroethylene by hybrid poplars," Environmental Science and Technology, vol. 31, no. 4, pp. 1062–1067, 1997. ··

145. P. V. R. Iyer, T. R. Rao, and P. D. Grover, Biomass Thermochemical Characterization Characterization, Indian Institute of Technology, Delhi, India, 3rd edition, 2002.

146. M.D. Hetland, J. R. Gallagher, D. J. Daly, D. J. Hassett, and L. V. Heebink, "Processing of plants used to phytoremediate lead-contaminated sites," A. Leeson, E. A. Forte, M. K. Banks, and V. S. Magar, Eds., pp. 129–136, Batelle Press.

147. C. D. Jadia and M. H. Fulekar, "Phytoremediation of heavy metals: recent techniques," African Journal of Biotechnology, vol. 8, no. 6, pp. 921–928, 2009.

148. USEPA, "Introduction to phytoremediation," Tech. Rep. EPA 600/R-99/107, United States Environmental Protection Agency, Office of Research and Development, Cincinnati, Ohio, USA, 2000.

In Situ Generated Colloid Transport of Cu and Zn in Reclaimed Mine Soil Profiles Associated with Biosolids Application

Jarrod O. Miller[1], Anastasios D. Karathanasis[2], and Christopher J. Matocha[2]

[1]ARS, USDA, 2611 W. Lucas St, Florence, SC 29501, USA

[2]N122 Ag Science North, Department of Plant and Soil Sciences, University of Kentucky, Lexington, KY 40546, USA

ABSTRACT

Areas reclaimed for agricultural uses following coal mining often receive biosolids applications to increase organic matter and fertility. Transport of heavy metals within these soils may be enhanced by the additional presence of biosolids colloids. Intact monoliths from reclaimed and undisturbed soils in Virginia and Kentucky were leached to observe Cu and Zn mobility with and without biosolids application Transport of Cu and Zn was observed in both solution and colloid associated phases in reclaimed and undisturbed forest soils, where the presence of unweathered spoil material and biosolids amendments contributed to higher metal release in solution fractions. Up to 81% of mobile Cu was associated with the colloid fraction, particularly when gibbsite was present, while only up to 18% of mobile Zn was associated with the colloid fraction. The colloid bound Cu was exchangeable by ammonium acetate, suggesting that it will release into groundwater resources.

INTRODUCTION

Water dispersible colloids may be a carrier vector for contaminants in the unsaturated soils zone, transporting metals to surface and groundwater [1–4]. The soil matrix is assumed to be a buffer to contaminant transport, due to its ability to sorb metals [2], but the mobilization of dispersible colloids from this matrix have been shown to transport contaminants [3, 4]. Reclaimed mine soils can be a source of heavy metals, released from unweathered spoil material, industrial wastes, fertilizers, power station fly ash, or biosolids applied during reclamation [5]. Copper (Cu), lead (Pb), or zinc (Zn) sulfides can leach from fresh spoil material [6], while cadmium (Cd), chromium (Cr), iron (Fe), manganese (Mn), and Pb can all be contained in phosphorus fertilizers [5].

Up to 95% of biosolids associated metals have been accounted for in the soil profile following biosolids application [7, 8], while under increasingly acidic conditions trace metals were observed to at least a 1 m depth in mine soils receiving biosolids [9]. Within this soil matrix, metal sorption is controlled by pH, clay mineralogy [10], or complexation with soil organic matter [11, 12]. It is commonly assumed that metals are adsorbed in the upper 15 to 30 cm of the soil matrix, thereby reducing their mobility [13, 14], but studies have observed significant metal transport by dispersed colloidal material [15, 16]. Therefore, early models which partition metals between an immobile solid and mobile liquid phase only have to be revised to include colloid particulate material as a third mobile solid phase, and a potential vector of metal transport [17].

Factors that affect colloid mobilization include clay mineralogy, ionic strength, pH, total clay content, soil moisture, and soil management [4]. Coal mining can destroy the original soil matrix, causing the loss of aggregation due to mining equipment and the oxidation of organic binding agents, both of which can increase mineral colloid release from the soil. Application of biosolids, a common reclamation procedure [18], may also be a source of organic colloids [19, 20]. Organic acids and humic material in the biosolids can chelate and bind metals, reducing, at least temporarily, their transport into groundwater [21]. Lime stabilized biosolids raise the soil solution pH, thus reducing metal solubility [18]. However, a basic pH can also cause organic colloids to be suspended into pore water, increasing the likelihood of being leached through the system [19].

Formation of pseudokarst channels is likely in reclaimed soils [6, 22, 23], and colloid transport through macropores can bypass impermeable spoil layers [3, 24]. On the other hand, high salt content is commonly associated with fresh mine spoils [6, 23], and increased ionic strength can aggregate colloids and reduce their mobility [17].

Due to their high surface area and charge density, colloids can be an important vector in transport of contaminants in the soil [25, 26]. Higher concentrations of Cd, Cu, and Zn within the dispersible clay fraction have been observed in soils receiving increasing rates of biosolids [8]. Colloid facilitated transport of dichlorodiphenyltrichloroethane (DDT), atrazine, and metals (Cu, Cr, Ni, Pb, Zn) have all been observed in packed and undisturbed columns [4]. Because of their affinity for pollutants, mobile colloids can also strip contaminants such as atrazine and zinc from the soil matrix [27]. Given that colloids can sorb metals from the soil matrix, it is likely that their presence will increase metal transport.

The objectives of this study were (1) to assess the mobility of Cu and Zn within reclaimed soils when spoil materials are placed beneath, (2) to assess the mobility of Cu and Zn within reclaimed soils receiving biosolids application, (3) compare metal mobilization to that occurring in similar undisturbed (natural) forest soils, and (4) to evaluate colloid, soil, and reclamation practices enhancing or inhibiting metal transport.

MATERIALS AND METHODS

Soil Monolith Preparation

Intact soil monoliths and disturbed material were obtained from the Powell River Project (PRP), near Wise, Virginia, in the southern Appalachian Mountains (30-year-old mine soils) and from Robinson Forest, near Jackson, Kentucky (5-year-old mine soils). This was done to observe any differences in colloid production as mine soils age. All reclaimed soils were from surface mined coal operations, where a top soil is replaced overtop of spoil materials. The monoliths were subjected to the following treatments for each study area (Figure 1); two replicated unmined forest soils, referred to as natural monoliths (VN), which were used as controls. Soils disturbed by coal mining constituted the reclaimed (VR), reclaimed soil + mine spoil material (VS), and reclaimed soil + mine spoil material + biosolids application treatments (VB). Kentucky treatments were natural monoliths (KN), reclaimed monoliths (KR), reclaimed soil and spoil (KS), while biosolids applied treatment (KB) constituted only of reclaimed soil + biosolids application, following low colloid elution from VB monoliths. This was done so that within Kentucky treatments, the effects of spoil and biosolids could be observed separately.

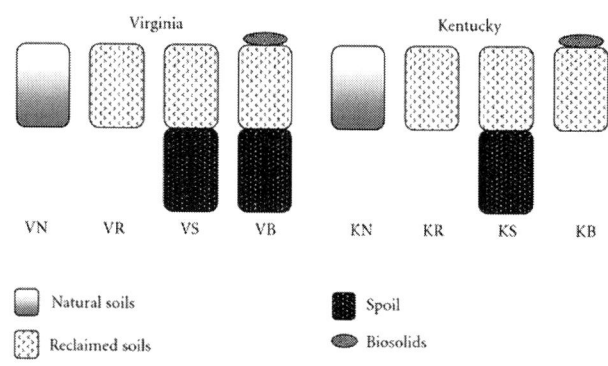

Figure 1: Diagram of the soil monoliths for the Virginia and Kentucky treatments.

The soil and disturbed monoliths (30 cm height by 18 cm diameter) were extracted from the Virginia study area by carving a pedestal, then trimming them with knives and soil picks to fit within a polyvinyl chloride (PVC) tube of 20 cm internal diameter and 30 cm height. The 2 cm gap between the PVC and the soil was sealed with expandable Poly-U-Foam (Kardol, 1-800-252-7365) to stabilize monoliths and prevent preferential flow along the walls. These intact monoliths preserve all of the structure and soil porosity present at the time of sampling. The PRP site in Virginia had been reclaimed by placing a spoil layer of siltstone and coal underneath a sandy soil material. These two layers were sampled separately as intact reclaimed and spoil monoliths, where each was 30 cm in height. Due to this, the reclaimed soil treatments were only 30 cm tall, while the reclaimed plus spoil treatments combined two separate 30 cm monoliths, for a total height of 60 cm (Figure 1). The soil and spoil materials were combined by placing the PVC encased monoliths flush with each other and sealing the edges with silicone to prevent leaking. Rock fragment content prevented consistently sized intact monoliths from being extracted from the reclaimed and spoils sites (for treatments KR, KS, and KB) from University of Kentucky's Robinson Forest. Instead, loose soil and spoil material were obtained to build monoliths in the lab. The reclaimed soils were obtained in two forms: (a) an intact, granular surface horizon of about 15 cm in thickness, and (b) a bucket of loose material from the subsurface. The material was taken back to the lab and combined into a 30 by 18 cm soil monolith. Wire mesh was made into a cylinder and placed into an empty PVC tube with a 20 cm internal diameter and 30 cm height. This mesh provided a mold for the loose soil and spoil material. Then a section of intact reclaimed surface horizon was placed on top to bring the total height to 30 cm. The 2 cm gap between the PVC and the soil was sealed with expandable Poly-U-Foam to stabilize monoliths and prevent preferential flow along the walls. The 30 cm tall spoil monolith was built in the same way for the KS treatment, but another 30 cm tall PVC column with wire mesh was placed above the spoil and Kentucky reclaimed soil material was placed above it, bringing the total height to 60 cm. Reclaimed soils receiving biosolids (KB) did not have a spoil monolith below them, while KN monoliths were obtained intact as described for the Virginia monoliths.

The lime stabilized biosolids material used in the study came from a local municipal wastewater treatment facility in Winchester (Clark

County), Kentucky. It was applied to the surface of the soil at a rate of 40 dry Mg ha^{-1}. A lower rate than normally applied (80 to 125 Mg ha^{-1}) was done due to the small surface area of the monoliths and the difficulty in mixing the biosolids into the surface of the monoliths without spilling over the sides. Instead, lower rates of biosolids were mixed by hand with the upper 6 inches of the reclaimed soils. The biosolids contained 7.0 g kg^{-1} P, 27.5 g kg^{-1} TKN, and had a calcium carbonate equivalent of 730 g kg^{-1} (Table 1).

Table 1: Selected chemical properties of the biosolids (processed by the EnviroData Group, Lexington, KY)

Biosolids characteristic	Value
pH	12.7
Percent Total Solids	41
g kg^{-1}	
Calcium carbonate equivalent (CCE)	730
Total Kjeldahl Nitrogen	27.5
Ammonia Nitrogen	1.2
Nitrate Nitrogen	0.03
Total P	7.0
Total K	1.1

Bulk Soil Cu and Zn Analysis

Natural, reclaimed, spoil, and biosolids materials were air-dried and passed through a 2 mm sieve. EPA method 3050b was used to extract Cu and Zn from 20 g of soil, spoil, or biosolids materials using HNO$_3$ and HCl and heating to 95°C. Extractants were analyzed by a Varian Vista Pro inductively coupled plasma analyzer (ICP) to determine preliminary levels of each metal in the materials. The pH and electrical conductivity (EC) were determined with a Denver Instrument Model 250 pH*ISE*conductivity meter. Ammonium acetate extracts were used to determine cation exchange capacity (CEC) and total exchangeable bases (TEB).

In Situ Colloid Elution

In situ colloid generation and elution from monoliths were assessed with leaching experiments using a rainfall simulator previously described by Miller et al. [28]. Deionized water (18 μS cm^{-1}) at a rate of 250 ml hour^{-1} (1.0 cm h^{-1}) was applied to the surface of each monolith with a peristaltic pump for approximately 2-3 pore volumes. Leaching was done in 6 hour pulses for Virginia monoliths and increased to 8 hour pulses for Kentucky monoliths. The cycle was repeated for 6 days until at least 2 pore volumes (pv) were achieved. The lower monolith boundary was kept at −10 cm using a Mariotte device.

Leachate was collected at the bottom of the monolith and tested for suspension concentrations, EC, pH, dissolved organic carbon (DOC), aromatic content of DOC [29], mineralogical composition of the colloids; and colloid particle size [28]. Larger particles were allowed to settle before sampling for suspended colloids. Mineralogical composition was performed on composite samples from each leaching cycle by X-ray diffraction (XRD) and thermo gravimetric analysis (TGA) using a Phillips PW 1840 diffractometer/PW 1729 X-ray generator and a TA 2000 thermogravimetric analyzer interfaced with a 951 DuPont TG module, respectively [30]. Colloidal particle size was determined on a Beckman Coulter N5 Submicron Particle Size analyzer on the first sample eluted from every cycle, if colloids were present. The software reports an average representative particle size for the sample.

Eluents were also tested for dissolved metals by taking a 50 mL aliquot from each hourly sample and passing it through a 0.2 μm filter to remove the colloidal material. Some colloids may be smaller than 0.2 um from these eluents, but observations of filtrates passing thru 0.2, 0.1 and 0.05 um filters showed no differences in colloid concentration. The filtered material was analyzed for dissolved metals by ICP. Following filtration, 20 mL of 1 M HCl/HNO$_3$ was passed through the same 0.2 μm filter containing the colloids to strip any bound metals. The HCl/HNO$_3$ filtrate was analyzed for metals by ICP and represents the colloid bound fraction.

RESULTS AND DISCUSSION

Metal Concentrations in Bulk Samples

Zinc concentrations in the digested samples ranged from 14.7 to 60.8 mg kg^{-1}, and Cu from 1.7 to 23.0 (Table 2). The highest recovered concentration of Zn was observed in the biosolids materials (60.8 mg kg^{-1}), while the highest Cu levels were extracted from the Kentucky spoil (23.0 mg kg^{-1}) and biosolids (21.4 mg kg^{-1}) materials, respectively. Biosolids may contain metals that are potential contaminants to the groundwater [31], but levels within these biosolids were well below EPA limits. Within the Virginia sites, Cu was highest in the spoil material, while Zn was highest in the reclaimed soils. The Kentucky spoil was recently exposed from a roadside cut at a surface mine, which probably explains the higher Cu and Zn concentrations observed within this material, possibly present as sulfate compounds. In contrast, the 30-year-old Virginia spoil material had roughly half the extractable Cu and Zn observed in the fresh Kentucky spoil. This indicates that the Virginia spoil may have released metals over time as it weathered, and we could expect the same with the Kentucky materials. Other properties of the bulk soil material have been discussed in Miller et al. [28, 32].

Table 2: Extractions of Cu and Zn in mg kg^{-1} by HCl/HNO$_3$ in soils, spoil, and biosolids

	VN^a	VR	VS	KN	KR	KS	Biosolids
	mg kg^{-1}						
Cu	1.7 (0.2)	4.3 (0.3)	11.3 (1.2)	4.9 (2.7)	2.4 (0.1)	23.0 (0.4)	21.4 (1.0)
Zn	14.7 (2.5)	23.6 (1.1)	16.5 (3.1)	19.1 (4.6)	17.8 (1.4)	49.8 (13.3)	60.8 (0.6)

α:VN: Virginia Natural, VR: Virginia Reclaimed, VS: Virginia Spoil, KN: Kentucky Natural, KR: Kentucky Reclaimed, KS: Kentucky Spoil.

*represents average value with standard deviation in parenthesis.

Cu and Zn Elution in Virginia Mine Soils

While VN monoliths had the largest mass of eluted colloids [32], they did not show a significant elution of colloidal Cu or Zn. The largest total (solution and colloid) mass of Cu was eluted from VB monoliths. However, there were no significant differences in the mass of soluble or colloidal Cu fractions in any of the Virginia treatments (Table 3). The VB monoliths also produced the highest mass of soluble (20.75 mg) and colloid-bound (0.78 mg) Zn fractions, while all other treatments were similar. The high initial breakthrough of dissolved Cu and Zn observed in this study is consistent with leachate properties observed in other mine soil profiles receiving biosolids [33–35].

Table 3: Total, colloidal, and percent-bound Cu and Zn for eluents from the Virginia and Kentucky treatments*

	VN[a]	VR	VS	VB	KN	KR	KS	KB
Total Cu (mg)	0.04 a (<0.01)	0.32 a (0.27)	0.04 a (0.05)	0.93 a (0.72)	0.12 b (0.04)	0.0 b (na)	0.49 b (0.10)	3.30 a (0.91)
Dissolved Cu (mg)	0.03 a (<0.01)	0.05 a (0.05)	0.04 a (0.05)	0.87 a (0.66)	0.02 b (<0.01)	0.0 b (na)	0.75 b (0.10)	3.04 a (0.87)
Colloid Cu (mg)	0.01 a (<0.01)	0.23 a (0.21)	0.01 a (<0.01)	0.07 a (0.06)	0.10 b (0.03)	0.0 c (na)	0.0 c (na)	0.26 a (0.04)
Bound Cu (%)	27.3 b (9.69)	82.0 a (1.64)	29.9 b (28.37)	6.5 b (1.18)	80.9 a (3.1)	0.0 c (na)	0.0 c (na)	8.2 b (1.08)
Total Zn (mg)	5.13 b (0.46)	3.29 b (2.60)	6.78 b (3.09)	21.53 a (8.53)	11.76 a (0.04)	2.75 a (1.47)	50.73 a (39.2)	7.04 a (6.8)
Dissolved Zn (mg)	5.09 b (0.45)	2.93 b (2.73)	6.74 b (3.10)	20.75 a (8.30)	11.44 a (0.08)	2.41 a (1.54)	50.73 a (39.2)	6.39 a (6.20)
Colloid Zn (mg)	0.04 b (<0.01)	0.36 b (0.13)	0.04 b (0.01)	0.78 a (0.22)	0.32 a (0.13)	0.34 a (0.07)	0.0 a (na)	0.65 a (0.60)
Bound Zn (%)	0.7 a (0.09)	18.2 a (18.4)	0.5 a (0.37)	3.7 a (0.46)	2.7 ab (1.05)	15.4 a (10.78)	0.0 b (na)	9.7 ab (0.78)

* = Statistical differences (LSD = 0.05) are represented by letters and are compared within row by sites only, standard deviations are in parenthesis, na = not applicable.

α = VN: Virginia Natural, VR: Virginia Reclaimed, VS: Virginia Spoil, VB: Virginia Biosolids, KN: Kentucky Natural, KR: Kentucky Reclaimed, KS: Kentucky Spoil, KB: Kentucky Biosolidsg.

The percentage of total Cu transport mediated by colloids (Table 3) was highest in the VR monoliths (82.0%), while all other treatments were similar. Eluents from VR monoliths also had the highest portion of colloid bound Zn (18.2%) compared to other Virginia treatments, but the differences were not significant. The higher Cu levels in the VR colloid fraction may be due to the mineralogical makeup of the eluted colloids [36, 37], which were dominated by 2:1 minerals and gibbsite [32]. Zinc eluted from VN monoliths was similar in total (solution + colloid) mass to VR and VS treatments, but less than 1% was colloid bound compared to 27.3% for Cu. Less than 4% of eluted Zn was transported in the VN, VS, and VB treatments; suggesting that Zn is dominantly transported in the dissolved fraction, unless gibbsite is present. The increased mobility of soluble metals may be attributed to the acidic eluents observed in VN, VS, and VB monoliths [9].

When spoil material was placed below the Virginia reclaimed monoliths (VS), colloid transport was significantly inhibited [32, 38], thus reducing the chance of colloid mediated transport of metals. Only trace amounts of total Cu (0.04 mg) were present in the eluent, with only up to 30% being colloid bound (Table 3). Within the VR eluents, Cu was dominantly transported in the colloid phase, but the restriction of colloid movement by a dense spoil reduced the overall total Cu mobility, with similar amounts of soluble Cu being released from VR and VS monoliths. Zinc, on the other hand, increased 2 fold in the dissolved phase when spoil was added to reclaimed soils, although extractable Zn was slightly lower in the spoil material. This may be due to the more acidic pH of the VS eluents [32], which allowed for greater soluble Zn mobility [38].

Concentrations of eluted colloids in VN, VR, and VS monoliths generally started off at their highest concentrations, and dropped throughout the leaching cycle [32]. Only in the VB monoliths was there a low but consistent release of colloidal material. Patterns of Cu and Zn elution (Figures 2(a) and 2(b)) were more erratic, with colloid associated metals showing spikes at several instances during leaching. A relatively large mass of Cu associated with VR colloids was eluted during an initial flushing stage, with another spike occurring at about 1 pv (Figure 2(a)). Colloids eluted from VB monoliths on the other hand, had several Cu spikes within 1 pv and another after 2 pv. This irregular elution pattern suggests that the variability in colloid particle size and mineralogy may make colloid mediated Cu transport difficult to predict.

Combining eluted soluble and colloid-bound Cu phases in Figures 3(a) and 3(b) does not contribute much more to the understanding of Cu mobility. Although it rose the overall total Cu mass eluted from VB monoliths, the spikes in concentration contributed by colloids remain, as with the VR and VN monoliths.

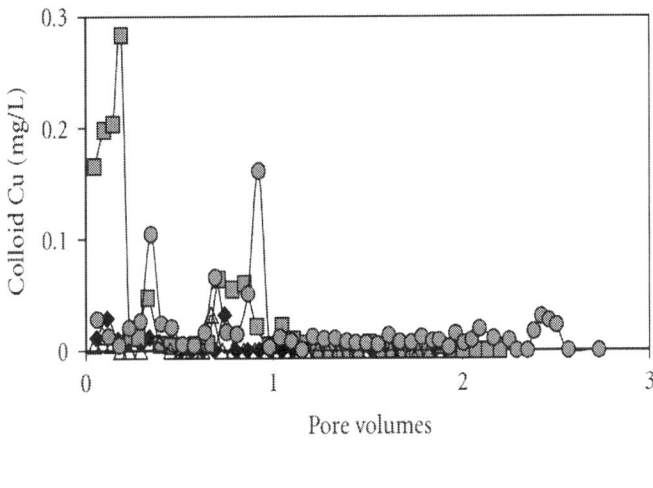

(a)

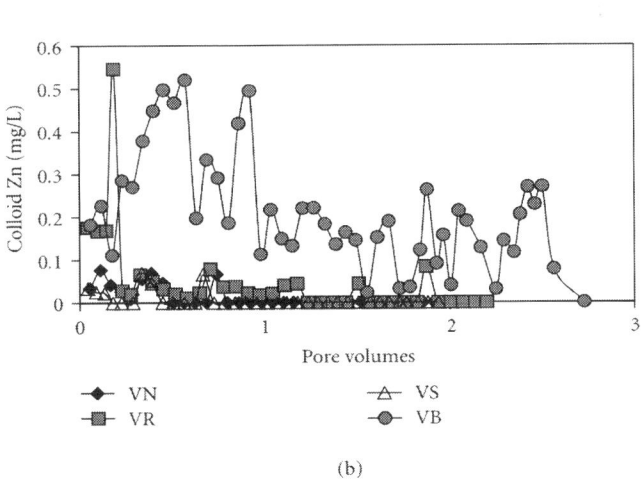

(b)

(b)

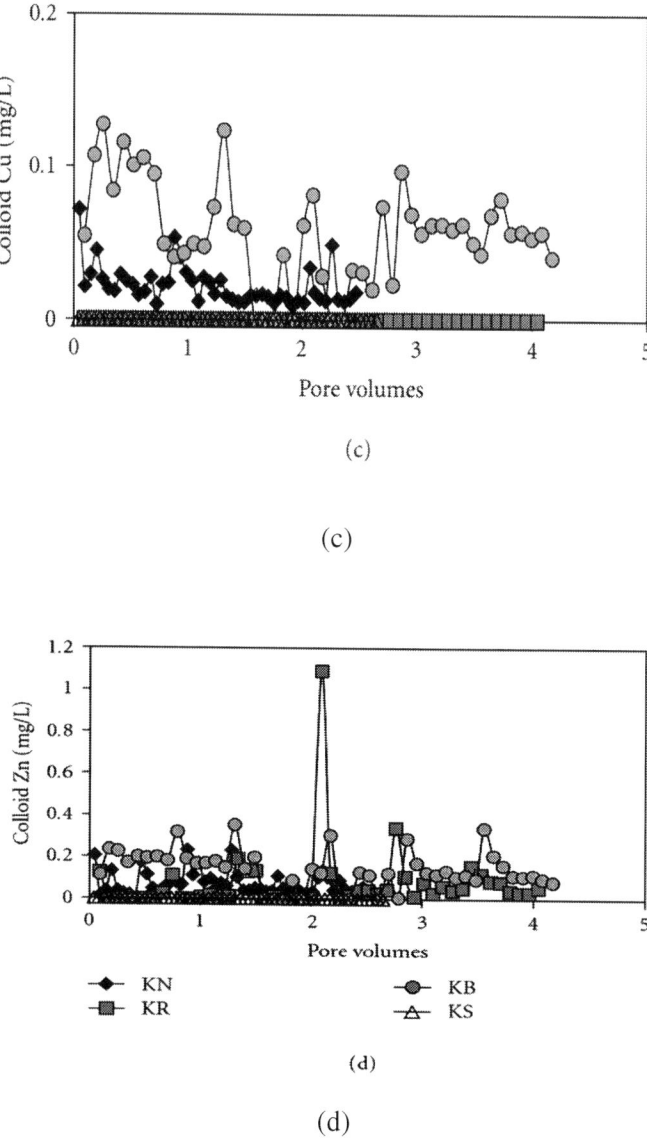

(c)

(c)

(d)

(d)

Figure 2: Colloid associated concentrations (mg L⁻¹) of Cu and Zn in Virginia (a, b) and Kentucky (c, d) monoliths over the entire leaching. VN: Virginia Natural, VR: Virginia Reclaimed, VS: Virginia Spoil, VB: Virginia Biosolids, KN: Kentucky Natural, KR: Kentucky Reclaimed, KS: Kentucky Spoil, KB: Kentucky Biosolids.

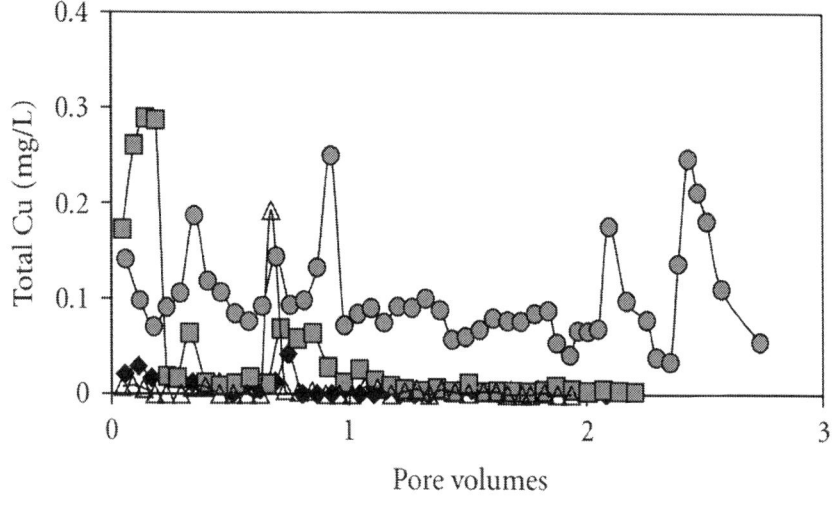

(a)

(a)

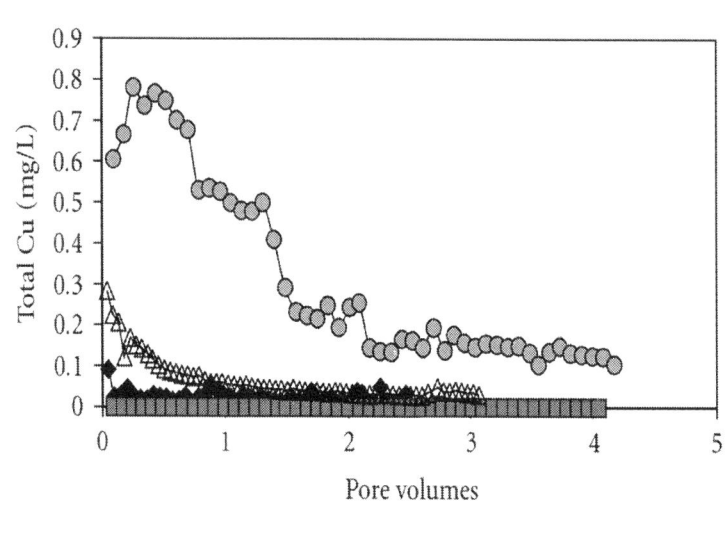

(c)

(b)

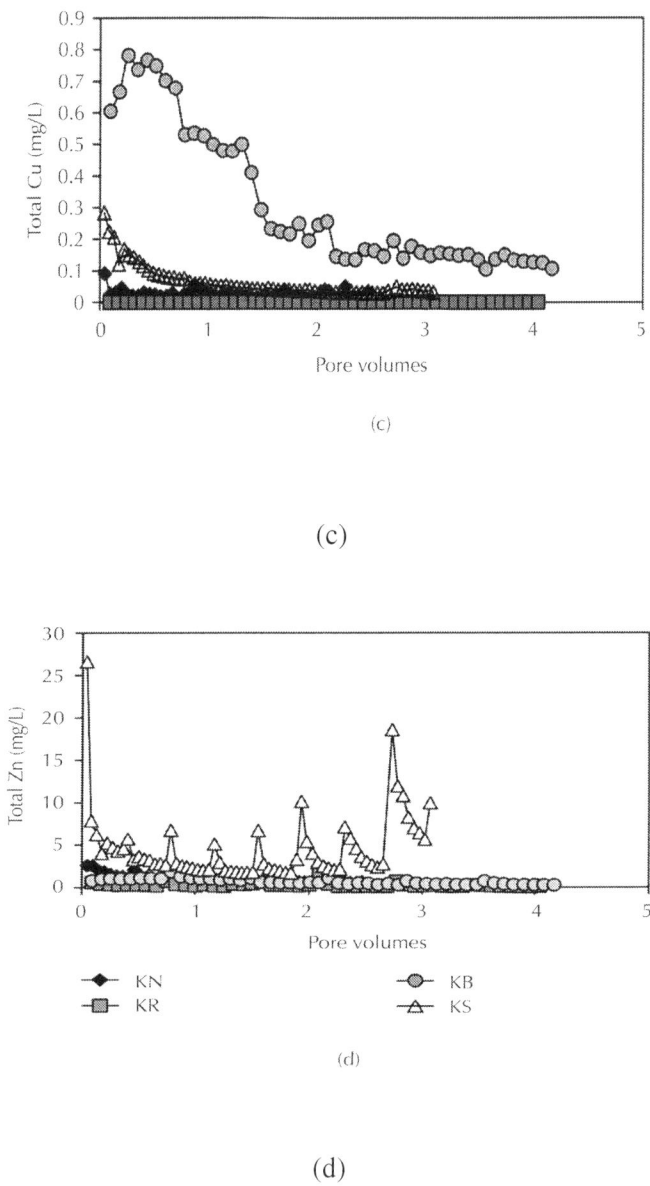

(c)

(c)

(d)

(d)

Figure 3: Total (colloid and solution) concentrations (mg L⁻¹) of Cu and Zn in Virginia (a, b) and Kentucky (c, d) monoliths over the entire leaching. VN: Virginia Natural, VR: Virginia Reclaimed, VS: Virginia Spoil, VB: Virginia Bio-solids, KN: Kentucky Natural, KR: Kentucky Reclaimed, KS: Kentucky Spoil, KB: Kentucky Biosolids.

Within the VN monoliths, total Zn was observed regularly in all eluent samples (Figure 3(b)), with most of the colloid fraction Zn eluted within 1 pv (Figure 2(b)), following colloid elution patterns [32]. Not many samples contained colloids [32] or bound metals after 1 pv in VR eluents (Figure 2(b)), but solution phase Zn was present in all samples (Figure 3(b)). The Zn concentration spikes observed in the VS breakthrough curves at the beginning of each leaching cycle are probably due to the diffusion of Zn from smaller pore spaces during periods of zero to low flow (Figure 3(b)). For biosolids amended treatments, the elution of soluble Zn (Figure 3(b)) was much smoother than the erratic behavior seen in Figure 2(b), except after 2 pv, where an upward shift in pH and colloid elution was observed [32]. The total mass of eluted Zn reached a peak concentration by 0.5 pv, (probably due to free and DOC complexed Zn released from the biosolids), tailing quickly thereafter (Figure 3(b)). While this elution pattern may be alarming for an initial flushing stage of metals to the groundwater following biosolids application, it may also suggest a quick depletion over repeated rainfall events [35].

Cu and Zn Elution in Kentucky Mine Soils

Eluents from KB monoliths had the greatest total, soluble, and colloid associated Cu, while all other Kentucky treatments were similar (Table 3). Increased Cu mobility has been previously observed in soil leachates due to surficial biosolids applications [33–35]. Within the Kentucky reclaimed soils (KR) no Cu was detected in either the soluble or colloid fractions, even though KR monoliths had the largest mass of colloids eluted and Cu was present in the bulk soil extracts (Table 2). The addition of unweathered spoil material (KS) beneath reclaimed monoliths resulted in detectable levels of Cu compared to reclaimed soils alone (KR). Colloid elution was not detected within 2 pv of leaching [28], so all reported Cu and Zn eluted from KS monoliths is from the solution phase.

Dissolved Zn increased 25 fold in KS eluents compared to other treatments, suggesting that the fresh unweathered spoil material contributed to Zn mobilization. The chemical properties of the eluents reflect only what it comes into contact with in the soil/spoil matrix [23], so the effects of toxic materials on pore water cannot be easily predicted by average carbonate and pyrite contents of spoils. In this case, the fresh spoil probably had unoxidized Zn minerals throughout the matrix, but if it had been mixed with other spoil types, Zn loads may have been lower. The application of biosolids also doubled Zn elution compared to reclaimed soils alone. Even the natural forest soil treatments (KN) had five times more Zn than KR eluents. This is probably associated with the leached sandy nature of the KR soils, due to their inherent characteristics as an Appalachian topsoil replacement. In spite of the large range in Zn elution, the high variability between the duplicated monoliths prevented significant statistical differences between the treatments.

The transport of soluble and colloidal Zn also doubled with biosolids application. An increase in DOC was observed in KB versus KR eluents [28], which may explain the increased transport of dissolved Cu and Zn. The greater mass of Zn associated with KB colloids, even with a lower colloid mass [28], indicates that these colloids carried a larger metal load. Colloids from KB monoliths may have more mineral-organic complexes due to the biosolids, thus increasing their ability to carry Cu and Zn. Overall; there was a greater cumulative release of Zn compared to Cu with biosolid application, which has been correlated with soil acidity in other mine soils receiving biosolids [9].

Although KN eluents had a low total Cu elution (0.12 mg), nearly 81% was bound to mobile colloids, greater than any other treatment. Kentucky reclaimed (KR) monoliths eluted no detectable Cu, while KS monoliths eluted no detectable colloids. The larger colloid bound transport of Cu in undisturbed forest soils may be due to the presence of colloidal gibbsite [28], suggesting that colloid mineralogical composition may be more influential than total colloid mass [36, 37]. Dissolved organic carbon was also higher in KN eluents, which may have influenced solution and colloidal transport of both Cu and Zn [28]. Cumulative Zn loads were the lowest in KR monoliths, but these soils also had the lowest extracted levels from bulk samples (Table 2).

Colloid associated leaching patterns of Cu and Zn for all four Kentucky treatments are shown in Figures 2(c) and 2(d). Again it can be observed that no colloidal Cu was detected within KR eluents, while no colloids were detected in KS eluents (Figure 2(c)). The pattern of colloid bound Cu from natural (KN) monoliths is erratic, often containing spikes in eluted colloidal Cu, which can be associated with pulses from flushing events [28]. Although colloids from KB (biosolids amended) monoliths typically carried greater Cu loads than KN colloids, at two sampling points Cu dropped below detection (Figure 1(c)), even though colloid concentration in KB eluents remained stable [28]. Colloids from KN monoliths, on the other hand, always had detectable amounts of bound Cu. This indicates that colloid mediated Cu transport may not be exclusively associated with high colloid concentrations, but may also be controlled by colloid mineralogy and particle size. When dissolved Cu is accounted for, elution patterns are much smoother, as larger concentrations of Cu in solution mask the variability in the colloid bound phase (Figure 3(c)). The addition of spoil material to reclaimed soils caused an initial flush of Cu elution, before a constant rate is observed. Therefore, fresh spoil material with unweathered Cu bearing minerals could be a long-term source of mobile Cu.

Similar results can be seen with Zn in Kentucky mine soils, where colloid bound Zn showed several spikes (Figure 2(d)). Some of these peaks between VN, VR, and VB monoliths occurred at the same point, and can be associated with initial pulses from the daily leaching cycle (Figure 2(d)). Other variations are more likely tied to differences in colloid mineralogy and particle size. These flushing cycle pulses can be better observed in Figure 3(d), where spikes in soluble Zn from KS eluents are evident for every initial leaching cycle samples.

Virginia versus Kentucky

Reclaimed soils in Virginia and Kentucky eluted a similar colloid bound and total mass of Zn, while Cu was not detectable in KR samples (Table 3). While KR monoliths produced almost 7 fold more colloids than their VR counterparts, total metal elution did not vary. This may indicate that reclaimed soils alone will not contribute to metal transport in solution or colloid phases. It is with the addition of spoil or biosolids that increases the potential for metal release in these systems.

The total mass of Zn was greater when spoil was added to KS treatments compared to VS treatments. Fresh spoil material in KS monoliths was also a more likely source of soluble Zn compared to the spoil that had undergone weathering for 30 years in VS monoliths. Colloid contribution to metal transport in both treatments with spoil was limited by bulk density and EC [28, 32], so the majority of this transport could be dominated by pH. Because most of the metal mobility within spoil amended monoliths was solution dominated, flushing patterns are more evident in the leaching cycles.

The addition of biosolids to reclaimed soils resulted in similar colloid bound metal transport patterns for both Zn and Cu (Figure 2). Release was largely unpredictable, with several peaks and valleys, but not often dropping below detection. This may indicate that soil water chemistry shifts may be moving through different pore sizes at varying rates, releasing colloids and metals in an unpredictable pattern. While the average pH of VB eluents is acidic, Miller et al. [32] reported sharp increases in sample pH following 2 pv of elution. The alkaline biosolids may release dissolved metals quickly, but it takes more time to overcome the pH buffer capacity of the soil. Cumulative mass Cu was 3 fold higher in KB eluents; probably due to the different overall length of the monoliths, which were 60 and 30 cm for VB and KB monoliths, respectively. While movement of metals associated with organic complexes below 80 cm has been observed with high loading rates [33], mobility beyond 30 cm appears to be limited with the lower application rates to these soils.

A comparison of natural forest soils from each study area does not yield similar results in colloid or metal elution. Total mass of colloids, Cu, and Zn were greater in KN eluents [28], indicating that undisturbed forest soils from different regions of Appalachia will vary in mobility, which should be expected. The ratio of total mass Zn:Cu was similar between the two forest soils, both being 100 fold higher in Zn and both dominantly showing greater Zn mobility in the dissolved phase.

Metal Associations

Selected colloid samples were treated with ammonium acetate () and 1 M HCl/HNO$_3$ to determine the exchangeable load of colloid bound metals (Table 4). The ammonium acetate method extracted Cu and Zn from all colloids, but the double acid extraction method exchanged significantly more Cu. This would verify the exchangeable nature of the Virginia and Kentucky eluted colloid bound metals (particularly Zn), and the increased potential for release to water resources.

Table 4: Colloid bound metals extracted by ammonium acetate (NH_4^+) and 1 M acid in mg L^{-1} in selected samples, with letters representing differences at 0.01 between the extraction methods

	Cu	Zn
	mg L^{-1}	
NH$_4^+$	0.003 b	0.261 a
HCl/HNO$_3$	0.017 a	0.266 a
% Exchangeable	17.6	98.1

CONCLUSIONS

The transport of Cu and Zn was observed in both the dissolved and colloid phases, with Zn being present in all treatments and Cu being detected in all but the KR eluents. The presence of unweathered spoil material and biosolids amendments contributed to higher metal release in soluble fractions, particularly Zn. The mobility of Cu was enhanced in the presence of gibbsite containing colloids but was less exchangeable than Zn. Therefore, Cu mobility was more limited and dependent on colloid movement, while dissolved Zn was ubiquitous in all systems. The total mass of eluted Zn in older reclaimed soils (VR, VS) was very similar, indicating that the contribution of Zn to pore waters in younger spoil materials (KS) will mitigate over time. Within the younger mine soils (KR); increases in surface organic material and plant uptake may also reduce metal leaching over time [39].

Except for the case of the younger spoil material, total metal concentrations were below 1 ppm and within drinking water standards. The total accumulation of metals was also not very high, considering the intensity of leaching in these lab experiments. In the case of the older, weathered forest soils, the presence of Cu and Zn in leachates indicates that natural release and leaching of these metals in the Appalachians must be low.

The application rate of biosolids used in the study was lower than the typical levels applied in the field and the concentration of the metals below the EPA limits. Therefore, our findings may underestimate actual field occurrences. However, our results indicate that an increase in transport of soluble metals directly after biosolids application is to be expected. Furthermore, mineral and organic colloid mobilization through larger diameter, saturated conduits occurring in cracks of disturbed soils may enhance transportability of larger metal loads. Whether these metal loads reach surface or groundwater will depend on the overall path length and the possibility of being adsorbed to the soil matrix. The addition of a dense spoil (VS) and salt laden fresh spoil (KS) limited colloid bound metal transport. This may also be the case in the field if saturated conduits are not present.

The large spatial and vertical matrix variability existing within mine land environments will make colloid contributions to mobility difficult though. Dispersion and movement of colloids cannot be retarded; therefore the application of biosolids to reclaimed lands should be offset by the depth to the water table, or runoff potential. Spoil materials which have the potential to release large amounts of salts should also be placed in upland positions, where longer path lengths to groundwater will reduce colloid and soluble metal loads.

ACKNOWLEDGMENTS

The authors would like to thank the Powell River Project in Wise, Va, and Yvonne Thomson for their assistance with this study.

REFERENCES

1. J. F. McCarthy and L. D. McKay, "Colloid transport in the subsurface. Past, present, and future challenges," Vadose Zone Journal, vol. 3, pp. 326–337, 2004.

2. J. M. Levin, J. S. Herman, G. M. Hornberger, and J. E. Saiers, "The effect of soil water tension on colloid generation within an unsaturated, intact soil core," in Colloids and Colloid-Facilitated Transport of Contaminants in Soils and Sediments, Foulum, pp. 107–111, Tjele, Denmark, October 2002, DIAS report.

3. A. K. Seta and A. D. Karathanasis, "Water dispersible colloids and factors influencing their dispersibility from soil aggregates," Geoderma, vol. 74, no. 3-4, pp. 255–266, 1996. ·

4. L. W. de Jonge, C. Kjaergaard, and P. Moldrup, "Colloids and colloid-facilitated transport of contaminants in soils: an

introduction," Vadose Zone Journal, vol. 3, pp. 321–325, 2004.

5. M. J. Haigh, "Soil quality standards for reclaimed coal-mine disturbed lands: a discussion paper,"International Journal of Surface Mining & Reclamation, vol. 9, no. 4, pp. 187–202, 1995.

6. G. Geidel and F. T. Caruccio, "Geochemical factors affecting coal mine drainage quality," inReclamation of Drastically Disturbed Lands, R. I. Barnhisel, W. L. Daniels, and R. Darmody, Eds., Agronomy Monograph 41, pp. 105–130, America Society of Agronomy, CSSA, and SSSA, Madison, Wis, USA, 2000.

7. S. P. McGrath and P. W. Lane, "An explanation for the apparent losses of metals in a long-term field experiment with sewage sludge," Environmental Pollution, vol. 60, no. 3-4, pp. 235–256, 1989.

8. B. F. Sukkariyah, G. Evanylo, L. Zelazny, and R. L. Chaney, "Recovery and distribution of biosolids-derived trace metals in a clay loam soil," Journal of Environmental Quality, vol. 34, no. 5, pp. 1843–1850, 2005. · ·

9. R. Stehouwer, R. L. Day, and K. E. Macneal, "Nutrient and trace element leaching following mine reclamation with biosolids," Journal of Environmental Quality, vol. 35, no. 4, pp. 1118–1126, 2006. · ·

10. N. Konig, P. Baccini, and B. Ulrich, "The influence of natural organic substances on the distribution of metals over soil and soil solution," Zeitschrift für Pflanzenernährung und Bodenkunde, vol. 149, pp. 69–82, 1986.

11. G. Sposito, L. J. Lund, and A. C. Chang, "Trace metal chemistry in arid-zone field soils amended with sewage sludge: I. Fractionation of Ni, Cu, Zn, Cd, and Pb in solid phases," Soil Science Society of America Journal, vol. 46, pp. 260–264, 1982.

12. A. A. Pohlman and J. G. McColl, "Kinetics of metal dissolution from forest soils by soluble organic acids," Journal of Environmental Quality, vol. 15, no. 1, pp. 86–92, 1986.

13. T. Streck and J. Richter, "Heavy metal displacement in a sandy soil at the field scale: I. Measurements and parameterization of sorption," Journal of Environmental Quality, vol. 26, no. 1, pp. 49–56, 1997.

14. L. Gove, C. M. Cooke, F. A. Nicholson, and A. J. Beck, "Movement of water and heavy metals (Zn, Cu, Pb and Ni) through sand and sandy loam amended with biosolids under steady-state hydrological conditions," Bioresource Technology, vol. 78, no. 2, pp. 171–179, 2001. · ·

15. J. F. McCarthy and J. M. Zachara, "Subsurface transport of contaminants: mobile colloids in the subsurface environment may alter the transport of contaminants," Environmental Science and Technology, vol. 23, pp. 496–502, 1989.

16. D. Grolimund, M. Borkovec, K. Barmettler, and H. Sticher, "Colloid-facilitated transport of strongly sorbing contaminants in natural porous media: a laboratory column study," Environmental Science and Technology, vol. 30, no. 10, pp. 3118–3123, 1996. · ·

17. D. Grolimund, K. Barmettler, and M. Borkovec, "Colloid facilitated transport in natural porous media: fundamental phenomena and modeling," in Colloidal Transport in Porous Media, F. Frimmel, F. von der Kammer, and H.-C. Flemming, Eds., pp. 3–24, Springer, New York, NY, USA, 2007.

18. K. C. Haering, W. L. Daniels, and S. E. Feagley, "Reclaiming mined lands with biosolids, manures, and papermill sludges," in Reclamation of Drastically Disturbed Lands, R. I. Barnhisel, W. L. Daniels, and R. Darmody, Eds., Agronomy Monograph 41, pp. 615–644, America Society of Agronomy, CSSA, and SSSA, Madison, Wis, USA, 2000.

19. A. D. Karathanasis and D. W. Ming, "Colloid-mediated transport of metals associated with lime-stabilized biosolids," Developments in Soil Science, vol. 28, pp. 49–62, 2002. ·

20. A. D. Karathanasis and D. M. C. Johnson, "Subsurface transport of Cd, Cr, and Mo mediated by biosolid colloids," Science of the Total Environment, vol. 354, no. 2-3, pp. 157–169, 2006. · · ·

21. W. E. Sopper, Municipal Sludge Use in Land Reclamation, Lewis Publishers, Boca Raton, Fla, USA, 1993.

22. T. A. Al and D. W. Blowes, "Storm-water hydrograph separation of run off from a mine-tailings impoundment formed by thickened tailings discharge at Kidd Creek, Timmins, Ontario," Journal of Hydrology, vol. 180, no. 1–4, pp. 55–78, 1996.

23. J. G. Skousen, A. Sexstone, and P. F. Ziemkiewicz, "Acid mine drainage control and treatment," inReclamation of Drastically Disturbed Lands, R. I. Barnhisel, W. L. Daniels, and R. Darmody, Eds., Agronomy Monograph 41, America Society of Agronomy, CSSA, and SSSA, Madison, Wis, USA, 2000.

24. J. F. McCarthy and L. Shevenell, "Processes controlling colloid composition in a fractured and karstic aquifer in eastern Tennessee, USA," Journal of Hydrology, vol. 206, no. 3-4, pp. 191–218, 1998. · ·

25. A. D. Karathanasis, "Subsurface migration of copper and zinc mediated by soil colloids," Soil Science Society of America Journal, vol. 63, no. 4, pp. 830–838, 1999.

26. P. M. Bertsch and J. C. Seaman, "Characterization of complex mineral assemblages: implications for contaminant transport and environmental remediation," Proceedings of the National Academy of Sciences of the United States of America, vol. 96, no. 7, pp. 3350–3357, 1999. · ·

27. C. D. Barton and A. D. Karathanasis, "Influence of soil colloids on the migration of atrazine and zinc through large soil monoliths," Water, Air, and Soil Pollution, vol. 143, no. 1–4, pp. 3–21, 2003. · ·

28. J. O. Miller, A. D. Karathanasis, O. O. Wendroth, C. J. Matocha, and C. D. Barton, "In situ colloid mobilization within biosolid amended soils following coal mine reclamation," in Proceedings of the NGWA/U.S. EPA Remediation of Abandoned Mine Lands Conference (#5019), Denver, Colo, USA, 2008.

29. S. J. Traina, J. Novak, and N. E. Smeck, "An ultraviolet absorbance method of estimating the percent aromatic carbon content of humic acids," Journal of Environmental Quality, vol. 19, no. 1, pp. 151–153, 1990.

30. A. D. Karathanasis and B. F. Hajek, "Revised methods for rapid quantitative determination of minerals in soil clays," Soil Science Society of America Journal, vol. 46, no. 2, pp. 419–425, 1982.

31. A. T. Lombardi and O. Garcia, "An evaluation into the potential of biological processing for the removal of metals from sewage sludges," Critical Reviews in Microbiology, vol. 25, no. 4, pp. 275–288, 1999.

32. J. O. Miller, A. D. Karathanasis, and O. O. B. Wendroth, "In situ colloid generation and transport in 30-year-old mine soil profiles receiving biosolids," International Journal of Mining, Reclamation and Environment, vol. 24, no. 2, pp. 95–108, 2010.

33. S. Brown, R. Chaney, and J. S. Angle, "Subsurface liming and metal movement in soils amended with lime-stabilized biosolids," Journal of Environmental Quality, vol. 26, no. 3, pp. 724–732, 1997.

34. M. B. McBride, B. K. Richards, T. Steenhuis, J. J. Russo, and S. Sauvé, "Mobility and solubility of toxic metals and nutrients in soil fifteen years after sludge application," Soil Science, vol. 162, no. 7, pp. 487–500, 1997.

35. M. B. McBride, B. K. Richards, T. Steenhuis, and G. Spiers, "Long-term leaching of trace elements in a heavily sludge-amended silly clay loam soil," Soil Science, vol. 164, no. 9, pp. 613–623, 1999.

36. F. A. Vega, E. F. Covelo, and M. L. Andrade, "Competitive sorption and desorption of heavy metals in mine soils: influence of mine soil characteristics," Journal of Colloid and Interface Science, vol. 298, no. 2, pp. 582–592, 2006. · · ·

37. F. M. Kishk and M. N. Hassan, "Sorption and desorption of copper by and from clay minerals," Plant and Soil, vol. 39, no. 3, pp. 497–505, 1973. · ·

38. S. A. Bradford, J. Simunek, M. Bettahar, M. T. van Genuchten, and S. R. Yates, "Significance of straining in colloid deposition: evidence and implications," Water Resources Research, vol. 42, no. 12, Article ID W12S15, 2006. ·

39. R. Krebs, S. K. Gupta, G. Furrer, and R. Schulin, "Solubility and plant uptake of metals with and without liming of sludge-amended soils," Journal of Environmental Quality, vol. 27, no. 1, pp. 18–23, 1998.

Bioremediation of Waters Contaminated with Heavy Metals Using Moringa oleifera Seeds as Biosorbent

Cleide S. T. Araújo[1], Dayene C. Carvalho[2],
Helen C. Rezende[2], Ione L. S. Almeida[2],
Luciana M. Coelho[3], Nívia M. M. Coelho[2], Thiago
L. Marques[2], and Vanessa N. Alves[2]

[1]State University of Goiás, Anápolis, GO, Brazil

[2]Institute of Chemistry, Federal University of Uberlândia, Uberlândia, MG, Brazil

[3]Department of Chemistry, Federal University of Goiás, Catalão, GO, Brazil

INTRODUCTION

Water is not only a resource, it is a life source. It is well established that water is important for life. Water is useful for several purposes including agricultural, industrial, household, recreational and environmental activities. Despite its extensive use, in most parts of the world water is a scarce resource. Ninety percent of the water on earth is seawater in the oceans, only three percent is fresh water and just over two thirds of this is frozen in glaciers and polar ice caps. The remaining unfrozen freshwater is found mainly as groundwater, with only a small fraction present above ground or in the air. Thus, almost all of the fresh water that is available for human use is either contained in soils and rocks below the surface, called groundwater, or in rivers and lakes.

The contamination of soil and water resources with environmentally harmful chemicals represents a problem of great concern not only in relation to the biota in the receiving environment, but also to humans. The continuing growth in industrialization and urbanization has led to the natural environment being exposed to ever increasing levels of toxic elements, such as heavy metals. Approximately 10% of the wastes produced by developed countries contain heavy metals. Figure 1gives some indication of the amounts of metal-containing waste produced in developed countries. Much of the discharge of metals to the environment comes from mining, followed by agriculture activities.

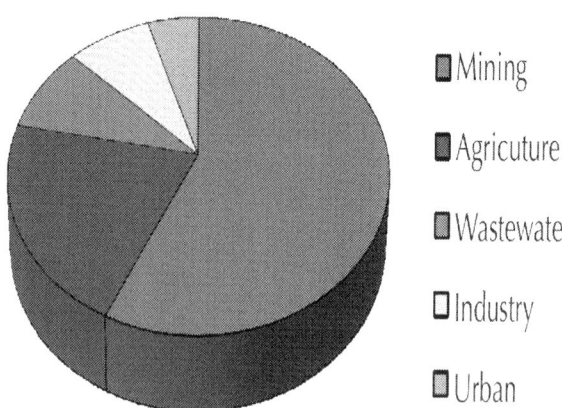

Figure 1: Waste containing heavy metals produced in developed countries [1].

Many different definitions have been proposed for heavy metals, some based on density, some on atomic number or atomic weight, and others on chemical properties or toxicity, which are not necessarily appropriate. For example, cobalt, iron, copper, manganese, molybdenium, vanadium, strontium and zinc are required to perform vital functions in the body and therefore cannot be considered as compounds with high toxicity or ecotoxic properties. Regarding the meaning of the term "heavy metal" it was found that there can be misinterpretation due to the contradictory definitions and lack of a coherent scientific basis [2].

In conventional usage "heavy" implies high density and "metal" refers to the pure element or an alloy of metallic elements. According to Duffus [2], a new classification should reflect our understanding of the chemical basis of toxicity and allow toxic effects to be predicted. Various publications have used the term "heavy metals" related to chemical hazards and this definition will also be used herein. Among the classes of contaminants, heavy metals deserve greater concern because of their high toxicity, accumulation and retention in the human body. Moreover, heavy metals do not degrade to harmless end products [3, 4]. It is well established that the presence of heavy metals in the environment, even in moderate concentrations, is responsible for producing a variety of illnesses of the central nervous system (manganese, mercury, lead, arsenic), the kidneys or liver (mercury, lead, cadmium, copper) and skin, bones, or teeth (nickel, cadmium, copper, chromium) [5].

Due to its properties, water is particularly vulnerable to contamination with heavy metals. Table 1 shows the maximum limits for some metals in drinking water, according to the US Environmental Protection Agency (US EPA) [6]. The US EPA requires that lead, cadmium and total chromium levels in drinking water do not to exceed 0.015, 0.005 and 0.1 mg L^{-1}, respectively. Corresponding values for other metals are presented in Table 1.

Table 1: Maximum acceptable concentrations of metals in drinking water according to the US EPA [6]

Element	US EPA Limit (mg L^{-1})
Antimony	0.006
Arsenic	0.010
Beryllium	0.004
Chromium (total)	0.1
Cadmium	0.005
Cupper	1.3
Lead	0.015
Mercury	0.002
Selenium	0.05
Silver	0.1

Within this context, and considering that heavy metals do not decay and are toxic even at low concentrations, it is necessary to remove them from various types of water samples. Of the conventional treatments used for the removal of metals from liquid waste, chemical precipitation and ion exchange are the predominant methods. However, they have some limitations since they are uneconomical and do not completely remove metal ions, and thus new removal processes are required [7-9]. Table 2 illustrates in more detail the advantages and limitations of the traditional methods applied to treat effluents.

Table 2: Traditional process used in wastewater treatment: advantages and disadvantages [10]

Process	Disadvantages	Advantages
Precipitation and filtration	For high concentrations	Simple
	Separation difficult	Low cost
	Not very effective	
	Produces sludge	

Biological oxidation and reduction	When biological systems are used the conversion rate is slow and susceptible to adverse weather conditions	Low cost
Chemical oxidation and reduction	Requires chemicals Applied to high concentrations Expensive	Mineralization Enables metal recovery
Reverse osmosis	High pressures Expensive	Pure effluent (for recycling)
Ion exchange	Responsive to the presence of particles Resins of high cost	Effective Enables metal recovery
Adsorption	Not effective for some metals	Conventional sorbents (coal)
Evaporation	Requires an energy source Expensive Produces sludge	Pure effluent obtained

For these reasons, alternative technologies that are practical, efficient and cost effective for low metal concentrations are being investigated. Biosorption in the removal of toxic heavy metals is especially suited as a 'nonpolluted' wastewater treatment step because it can produce close to drinking water quality from initial metal concentrations of 1-100 mg L^{-1}, providing final concentrations of < 0.01-0.1 mg L^{-1} [11]. Biosorption has been defined as the ability of certain biomolecules or types of biomass to bind and concentrate selected ions or other molecules from aqueous solutions. It should to be distinguished from bioaccumulation which is based on active metabolic transport; biosorption by dead biomass is a passive process based mainly on the affinity between the biosorbent and the sorbate [12]. The biosorption of heavy metals by non-living biomass of plant origin is an innovative and alternative technology for the removal of these pollutants from aqueous solution and offers several advantages such as low-cost biosorbents, high efficiency, minimization of chemical and/or biological sludge, and regeneration of the biosorbent [13].

Recently, natural adsorbents have been proposed for removing metal ions due to their good adsorption capacity. Technologies based on the use of such materials offer a good alternative to conventional technologies for metal recovery. In this context, *Moringa oleifera* represents an alternative material for this purpose [14-16].

MORINGA OLEIFERA

Moringa oleifera is the best known species of the *Moringaceae* family. *Moringaceae* is a family of plants belonging to the order *Brassicales*. It is represented by fourteen species and a single genus (*Moringa*), being considered an angiosperm plant. It is a shrub or small tree which is fast growing, reaching 12 meters in height. It has an open crown and usually a single trunk (Figure 2). It grows mainly in the semi-arid tropics and subtropics. Since its preferred habitat is dry sandy soil, it tolerates poor soils, such as those in coastal areas [17].

Figure 2: Tree of *Moringa oleifera* species [18].

Native to northern India, it currently grows in many regions including Africa, Arabia, Southeast Asia, the Pacific and Caribbean Islands and South America [3, 16, 19]. It is cultivated for its food, medicinal and culinary value and its leaves, fruits and roots are the parts used. It is commonly known as the 'horseradish' tree arising from the taste of a condiment prepared from the roots or 'drumstick' tree due to the shape of the pods. Figures 3 and 4 show the pods and seeds of this tree. *M. oleifera* has a host of other country-specific vernacular names, an indication of the significance of the tree around the world [16, 20-23].

Figure 3: Pods of *Moringa oleifera* [18].

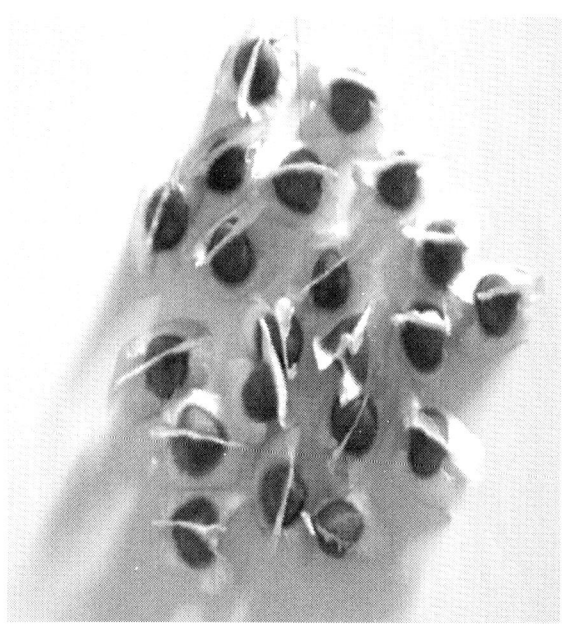

Figure 4: Seeds of *Moringa oleifera* [18].

Research has focused on the use of *M. oleifera* seeds and fruits in water purification and the treatment of turbid water is the best-known application. The seeds of various species contain cationic polyelectrolytes which have proved to be effective in the treatment of water, as a substitute for aluminum sulfate. Interest in the study of natural coagulants for water clarification is not new. The coagulant is obtained from a byproduct of oil extraction and the residue can be used as a fertilizer or processed for animal fodder. Compared to the commonly used coagulant chemicals, *Moringa oleifera* has a number of advantages including low cost, biodegradable sludge production and lower sludge volume, and also it does not affect the pH of the water. Apart from turbidity removal, *M. oleifera* seeds also possess antimicrobial properties [24, 25], although the mechanism by which seeds act upon microorganisms is not yet fully understood.

Tissues of *M. oleifera* from a wide variety of sources have been analyzed for glucosinolates and phenolics (flavonoids, anthocyanins, proanthocyanidins, and cinnamates). *M. oleifera* seeds reportedly contain 4-(-L-rhamnopyranosyloxy)-benzylglucosinolate in high concentrations. Roots of *M. oleifera* have high concentrations of both 4-(-L-rhamnopyranosyloxy)-benzylglucosinolate and benzyl glucosinolate. Leaves contain 4-(-L-rhamnopyranosyloxy)-benzylglucosinolate and three monoacetyl isomers of this glucosinolate and only 4-(-L-rhamnopyranosyloxy)-benzylglucosinolate has been detected in *M. oleifera* bark tissue [26]. Every glucosinolate contains a central carbon atom which is bonded to the thioglucose group (forming a sulfated ketoxime) via a sulfur atom and to a sulfate group via a nitrogen atom. These functional groups containing sulfur and nitrogen are good metal sequesters from aqueous solution. The leaves of *M. oleifera* reportedly contain quercetin-3-*O*-glucoside and quercetin-3-*O*-(6''-malonyl-glucoside), and lower amounts of kaempferol-3-*O*-glucoside and kaempferol-3-*O*-(6''-malonyl-glucoside), along with 3-caffeoylquinic acid and 5-caffeoylquinic acid. Neither proanthocyanidins nor anthocyanins have been detected in any of the tissues [26]. Although*M. oleifera* seeds have been most widely applied as a coagulant agent, many studies have been performed in order to explore other potential applications of this material, especially in the removal of metals from aqueous systems.

BIOSORPTION OF METALS USING MORINGA OLEIFERA

Since *Moringa oleifera* seeds have the ability to retain metals, it is necessary to define and to understand the functional groups responsible for the adsorption phenomenon. Biosorption by dead biomass or by some molecules and/or their active groups is a passive process based mainly on the affinity between the biosorbent and the sorbate. In this case, the metal is sequestered by chemical sites naturally present in the biomass. The diagram in Figure 5 illustrates the main steps in this process. In most cases, the biosorption process is rapid and takes place under normal temperature and pressure. After the process of phase separation a biomass "charged" with metal ions and an effluent free of contamination are obtained. Two paths can be followed to deal with the "contaminated" biomass, the one of greatest interest being biosorbent regeneration and metal recovery. This process is the most attractive because biomass can be used for the removal of other metal species from other contaminated effluents. The other option is the destruction of the biomass, which offers no possibility of reuse.

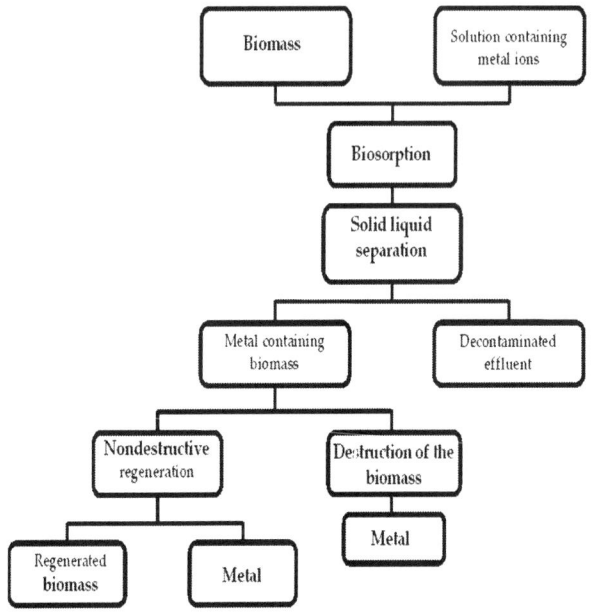

Figure 5: Main steps in biosorption process [27].

The mechanisms associated with heavy metal biosorption by biomass are still not clear; however, it is important to note that this process is not based on a single mechanism. Since metals may be present in the aquatic environment in dissolved or particulate forms, they can be dissolved as free hydrated ions or as complex ions chelated with inorganic ligands, such as hydroxide, chloride or carbonate, or they may be complexed with organic ligands such as amines, humic or fulvic acids and proteins. Metal sequestration occurs through complex mechanisms, including ion-exchange and complexation, and it is quite possible that at least some of these mechanisms act simultaneously to varying degrees depending on the biomass, the metal ion and the solution environment.

In reference [28] indicated that ion-exchange is an important concept in biosorption, because it explains many of the observations made during heavy metal uptake experiments. In this context, the term ion-exchange does not explicitly identify the mechanism of heavy metal binding to biomass, and electrostatic or London–van der Waals forces should be considered as the precise mechanism of chemical binding, i.e., ionic and covalent bonds. Figure 6 provides a schematic representation of an ion-exchange mechanism for a biosorbent material where "Me" represents a metal with valence +2.

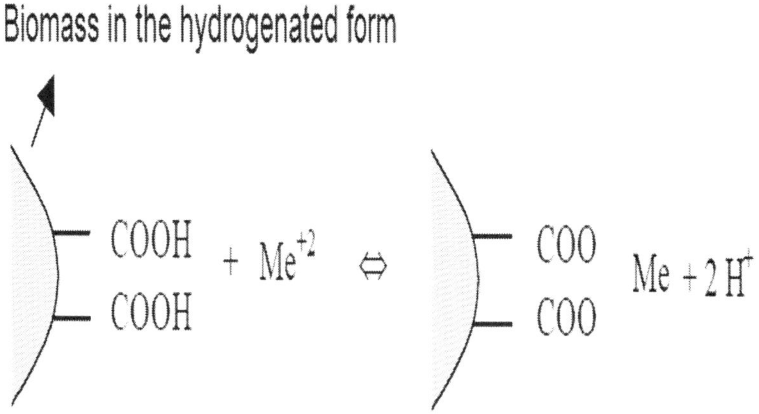

Biomass in the hydrogenated form

Figure 6: Schematic diagram of an ion exchange mechanism [29].

The seeds of *Moringa oleifera* and its parts can be classified as lignocellulosic adsorbents, consisting mainly of cellulose, hemicellulose and lignin. These functional groups are comprised of macromolecules that have the ability to absorb metal ions through ion exchange or complexation [30] phenomena which occur on the surface of the material through the interaction of the metal with the functional groups present. In order to understand the adsorption process it is also important to characterize the biomass material. Several techniques can be used to define the functional groups responsible for the adsorption phenomenon.

Infrared spectroscopy is an important technique in the qualitative analysis of organic compounds, widely used in the areas of natural products, organic synthesis and transformations. It is applied as a tool to elucidate the functional groups which may be present in substances [31], particularly with respect to the availability of the main groups involved in adsorption phenomena.

Figure 7 shows FT-IR spectra for *Moringa oleifera* seeds which verify the presence of many functional groups, indicating the complex nature of this material. The bandwidth centered at 3420 cm^{-1} may be attributed to the stretching of OH bonds present in proteins, fatty acids, carbohydrates and lignin units [32]. Due to the high content of protein present in the seed there is also a contribution in this region from N-H stretching of the amide bond. The peaks present at 2923 cm⁻¹ and 2852 cm⁻¹, respectively, correspond to asymmetric and symmetric stretching of the C-H bond of the CH$_2$ group. Due to the high intensity of these bands it is possible to assign them to the predominantly lipid component of the seed, which is present in a high proportion similar to that of protein [33]. In the region of 1800-1500 cm⁻¹ a number of overlapping bands are observed and between 1750 and 1630 cm⁻¹ this can be attributed to C=O stretching. Due to the heterogeneous nature of the seed, the carbonyl group may be bonded to different neighborhoods as part of the fatty acids of the lipid portion or amides of the protein portion. The carbonyl component that appears due to the presence of lipids can be seen at 1740 and 1715 cm⁻¹, as can be observed in the infrared spectra as small peaks, and the shoulders forming part of the main band that appears at 1658 cm⁻¹ are attributed to the carbonyl amides present in the protein portion. The peak observed at 1587 cm⁻¹ may be attributed to stretching connecting CN and also the deformation of the N-H bond present in the proteins of seeds [34, 35].

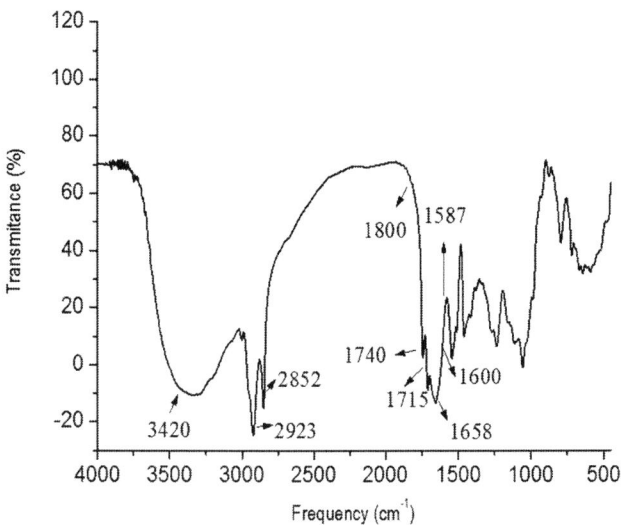

Figure 7: FT-IR spectrum of *Moringa oleifera* seeds. The arrows indicate the maximum signal obtained [36].

Among the various techniques for material characterization, the X-ray diffraction (XRD) technique is recommended for the evaluation of the presence of crystalline phases present in natural materials. In general, we can classify materials as amorphous, semicrystalline or crystalline. Figure 8 shows the XRD patterns for *M. oleifera* seeds. The XRD pattern for crushed seeds, due to the high amount of oils and proteins present in the composition of the material which represent around 69% of the total mass [36], shows unresolved signals (predominantly amorphous). For this reason intact seeds are analyzed, constituting a complex matrix comprised of a wide variation of substances including proteins, lipid structures and, to a lesser extent, carbohydrates. It was possible to separate a broad peak at around 2 equals 10°. The presence of this peak is probably associated with the diffraction of the protein constituent surrounded by other components which have a more amorphous pattern [37]. The amorphous nature of the biosorbent suggests that the metal ion could more easily penetrate the biosorbent surface.

Thermogravimetric (TG) analysis was used to characterize the decomposition stages and thermal stability determined through the mass loss of a substance subjected to a constant heating rate for a specified time. The mass loss curve for a sample of *Moringa oleifera* seeds can be observed in Figure 9, showing a typical profile that indicates several stages of the decomposition process. This thermogravimetric curve verifies the sample heterogeneity, since the intermediates formed are a mixture of several components. The mass loss curve can be divided into three stages: i) the first step occurs from 30°C to 128°C where a mass loss in the order of 8%, associated with water desorption, was observed. The amount of water loss from seeds determined by this technique is similar to the value of 8.9% found in [38]; ii) in the second step 32% of mass loss was observed in the temperature range of 128–268°C. This stage occurs due to the decomposition of organic matter, probably the protein component, present in seeds; and iii) the third step occurs from 268°C to 541°C with decomposition of the greater part of the seed components, which probably includes fatty acids, for example, oleic acid has a boiling point of 360°C. At 950°C a total residue of around 14.6% was observed, due to the ash content and probably inorganic oxides.

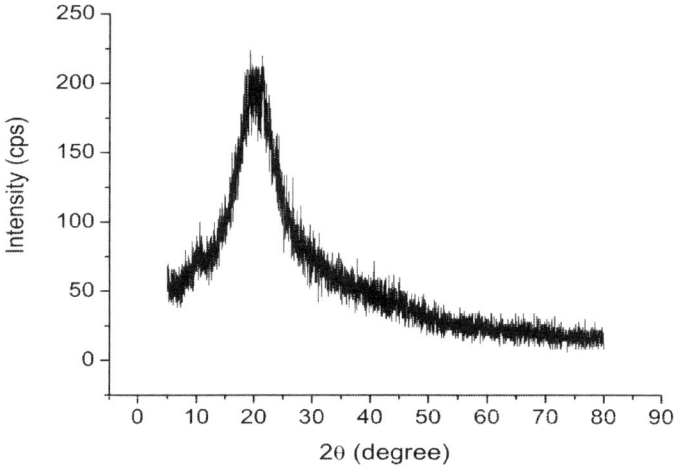

Figure 8: X–ray diffractogram for *Moringa oleifera* seeds [36].

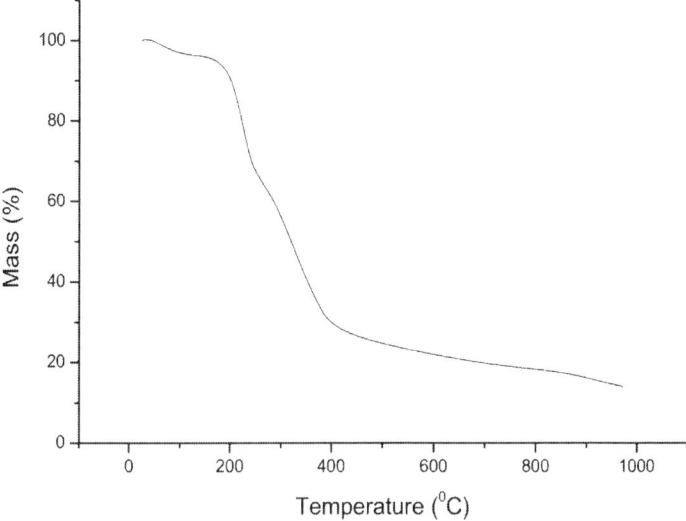

Figure 9: Thermogravimetric curve for *Moringa oleifera* seeds [36].

The morphological characteristics of the crushed seeds obtained using a scanning electron microscopy (SEM) can be seen in Figure 10. The results reveal that the material exhibits a relatively porous matrix with heterogeneous pore distribution. This feature is attributed to the fact that the whole seed comprises a wide variety of biomass components. The presence of some deformations on the surface of the plant tissue can be observed, containing available sites, from which it is possible to infer that the adsorbent provides favorable conditions for the adsorption of metal species in the interstices [35].

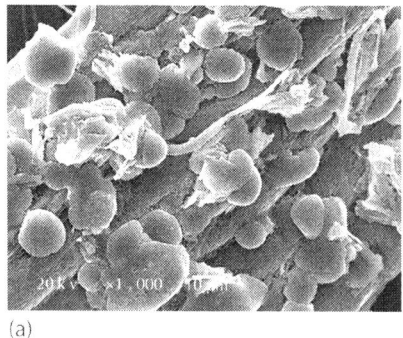
(a)

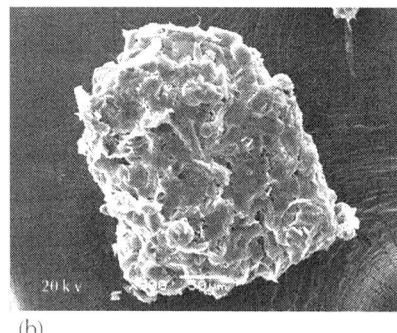
(b)

Figure 10: Scanning electron micrographs of *Moringa oleifera*. In the order of (a) 10 μm and (b) 50 μm [36].

INFLUENCE OF PARAMETERS IN BIOSORPTION PROCESS

Many variables can influence metal biosorption and experimental parameters such as temperature, stirring time, pH, particle size of the biomass, ionic strength and competition between metal ions can have a significant effect on metal binding to biomass. The biomass mass also influences the adsorption process because as the adsorbent dose increases the number of adsorbent particles also increases and there is greater availability of sites for adsorption. Some of the most important factors affecting metal binding are discussed below. In general, adsorption experiments are carried out in batch mode.

The pH is one of the most important parameters affecting any adsorption process. This dependence is closely related to the acid-base properties of various functional groups on the adsorbent surfaces [39]. The literature shows that a heterogeneous aqueous mixture of *M. oleifera* seeds contains various functional groups, mainly amino and acids groups. These groups have the ability to interact with metal ions, depending on the pH. An increase in metal adsorption with increasing pH values can be explained on the basis of competition between the proton and metal ions for the same functional groups, and a decrease in the positive surface charge, which results in a higher electrostatic attraction between the biosorbent surface and the metal [40]. Low pH conditions allow hydrogen and hydronium ions to compete with metal binding sites on the biomass, leading to poor uptake. Biosorbent materials primarily contain weak acidic and basic functional groups. It follows from the theory of acid–base equilibrium that, in the pH range of 2.5–5, the binding of heavy metal cations is determined primarily by the dissociation state of the weak acidic groups. Carboxyl groups (–COOH) are important groups for metal uptake by biological materials. At higher solution pH, the solubility of a metal complex decreases sufficiently for its precipitation, leading to a reduced sorption capacity. Therefore, it is recommendable to study biosorption at pH values where precipitation does not occur. Biomasses are materials with an amphoteric character; thus, depending on the pH of the solution, their surfaces can be positively or negatively charged. At pH values greater than the point of zero discharge (*pHpzc*), the biomass surface becomes negatively charged, favoring the adsorption of cationic species. However, adsorption of anionic species will be favored at pH < *pHpzc*. The *pHpzc* of the *M. oleifera* seeds is between 6.0 and 7.0 [41], indicating that the surface of the biosorbent presents acid characteristics.Figure 11 illustrates the surface charge or the point of zero net proton charge of *Moringa oleifera*seeds. The surface charge of the seeds is positive at pH < PZC, is neutral at pH = PZC and is negative at pH > PZC. The variation in pH caused by protonation and deprotonation of the adsorbent reflects the presence of functional groups. Table 3 shows the use of components of the *M oleifera* in the pH range of 2.5 to 8.0.

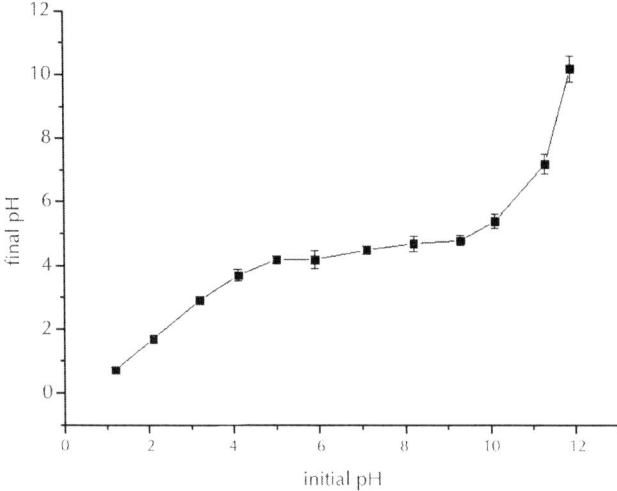

Figure 11: Point of zero net proton charge of *Moringa oleifera* seeds.

It has been noted that the temperature can influence the sorption process. Simple physical sorption processes are generally exothermic, i.e., the equilibrium constant decreases with increasing temperature. According to data reported in the literature (Table 3), the binding of the metal to different parts of the *M. oleifera* plant can be observed when the temperature is raised from 22 to 50 °C.

The contact time (or stirring time) is another important parameter that influences the efficiency of the adsorption process. As can be seen in Table 3, a period of 5 min was chosen for the nickel sorption process and good results were obtained; however, longer times (240 min) are required when using activated carbon.

Moringa oleifera is capable of directly sorbing metal ionic species from aqueous solutions. An interesting characteristic assigned to these biosorbents is the high abrasive content and the relative chemical resistance, allowing them to be subjected to different chemical treatments to increase their affinity and/or specificity for metal ions. Results previously published show the potential use of untreated seeds, although biosorbent materials are generally derived from plant biomass through different kinds of simple procedures. They may be chemically pretreated for better performance and/or suitability for process applications. However, good results have been obtained when the seeds were treated with NaOH. This treatment can remove organic and inorganic matter from the sorbent surface. Chemical treatments are commonly performed employing alkaline solutions or with phosphoric and citric acids [42]. Recently, however, efforts have been made to remove and subsequently also recover metals. Metal-saturated biosorbent materials can be easily regenerated applying a simple (e.g. acidic) wash which then contains a very high concentration of released metals in a small volume, making the solution quite amenable to metal recovery.

Table 3: Study parameters for the removal of metal ions using *Moringa oleifera*

Moringa Oleifera	Modifying agent(s)	Heavy metal	Temperature (°C)	pH	Contact time (min)	Ref.
Seeds	Petroleum ether	Cd (II) Cu (II) Co (II) Ni (II) Pb (II)	22	3.5 – 8.0	60	[4]
Leaves	NaOH and Citric acid	Cd (II) Cu(II) Ni(II)	40	5.0	50	[32]
Bark	Original state	Ni(II)	50	6.0	60	[35]
Wood	Activated carbon	Cu(II) Ni(II) Zn(II)	30	6.0	240	[31]
Leaves	NaOH and Citric acid	Pb(II)	40	5.0	50	[34]
Bark	Original state	Pb(II)	25	5.0	30	[19]

Pod	Original state	Zn(II)	30	7.0	50	[16]
	NaOH,					
	H_2SO_4					
	CTAB					
	HCl					
	$Ca(OH)_2$					
	Triton X-100					
	H_3PO_4					
	$Al(OH)_3$					
	SDS					
Shelled seeds	Original state	Cd(II)	-	6.5	40	[15]
		Cr(III)		6.5		
		Ni(II)		7.5		
Shells	Original state	As (III)	-	7.5	60	[43]
		As (V)		2.5		
Husk and pods	Unmodified	Pb(II)	30	5.8	120	[3]
	CTAB					
	H_3PO_4					
	H_2SO_4					
	HCl					
Shelled seeds	Original state	Cd(II)	-	6.5	40	[14]
Seeds	Original state	Ag(I)	25	6.5	20	[36]
Seeds	NaOH	Ni(II)	25	4.0-6.0	5	[44]

[i] - CTAB: Cetyl trimethylammonium bromide, SDS: Sodium dodecyl sulfate

ADSORPTION MODELS

An important physicochemical aspect in terms of the evaluation of sorption processes is the sorption equilibrium. Adsorption isotherms are a basic requirement in understanding how the adsorbate is distributed between the liquid and solid phases when the adsorption process reaches the equilibrium state [45, 46]. Over the years a wide variety of isotherm models have been introduced. The most commonly used isotherm models include Langmuir [47], Freundlich [48], Dubinin-Radushkevich [49] and Temkin [50].

It can be observed that in most of the cases the Langmuir adsorption model has been successfully used to predict metal adsorption processes. The Langmuir isotherm model assumes monolayer adsorption onto an adsorbent surface containing a finite number of identical sites and without interaction between adsorbed molecules. The Langmuir isotherm model assumes that: each site can accommodate only one molecule or atom; the surface is energetically homogenous; there is no interaction between neighboring adsorbed molecules or atoms; and there are no phase transitions [51]. The Langmuir equation is expressed as follows:

$$q_e = \frac{q_m K_L C_e}{1 + K_L C_e}$$

(1)

where q_e is the amount of metal adsorbed at equilibrium (mg g^{-1}), C_e is the concentration of metal in solution at equilibrium (mg L^{-1}), and q_m (mg g^{-1}) and KL (L mg^{-1}) are the Langmuir constants related to the adsorption capacity (amount of adsorbate needed to form a complete monolayer) and adsorption energy, respectively. The constants q_m and K_L can be calculated from the intercepts and the slopes of the linear plots of C_e/q_e versus C_e.

The Freundlich model describes adsorption onto an energetically heterogeneous surface not limited by the monolayer capacity [48]. It can be presented in the following form:

$$q_e = K_f C_e^n$$

(2)

where q_e is the amount of metal adsorbed at equilibrium (mg g^{-1}), C_e is the concentration of metal in solution at equilibrium (mg L^{-1}), and K_f (mg g^{-1})(L mg)$^{(1/n)}$ and n (g L^{-1}) are the Freundlich constants related to the multilayer adsorption capacity and adsorption intensity, respectively. According to the theory, n values between 1 and 10 represent favorable adsorption conditions [52]. Values of K_f and n can be calculated from the slope and intercept of the plot of Log q_e versus Log C_e. Experimental adsorption results with high coefficient correlation (R^2) values obtained for Freundlich isotherms have been reported as shown in Table 4.

The Dubinin-Radushkevich model has been used to distinguish between physical and chemical adsorption [53]. The Dubinin-Radushkevich is more general than the Langmuir model because it does not assume a homogenous surface or constant sorption. The Dubinin-Radushkevich equation is given by:

$$q_e = q_m e^{(-\beta \in^2)}$$

(3)

where, q_m (mg g^{-1}) is the theoretical sorption capacity (mol g^{-1}), ε is the Polanyi potential which is related to the equilibrium concentration and the constant gives the mean energy of sorption, E (KJ mol^{-1}). The constants q_m and β are obtained from the intercept and slope of ln q_e versus ε^2, respectively. If the magnitude of E is between 8 and 16 KJ mol^{-1} the adsorption process proceeds by ion-exchange or chemisorptions, while for values of E < 8 KJ mol^{-1} the adsorption process is of a physical nature [54]. In [31] reported that the sorption energy (E) values obtained with the Dubinin-Radushkevich model showed that the interaction between metal ions and the adsorbent proceeded principally by physical adsorption.

The Temkin isotherm model is based on the assumption that the heat of adsorption of all the molecules in the layer decreases linearly with coverage due to adsorbent-adsorbate interactions, and the adsorption is characterized by a uniform distribution of binding energies, up to a maximum binding energy [50]. The model is represented by the following equation:

$$q_e = B \ ln \left(AC_e \right)$$

(4)

where A $(L\,g^{-1})$ and B $(J\,mg^{-1})$ are Temkin isotherm constants relating to adsorption potential and heat of adsorption, respectively. A plot of q_e versus ln C_e gives the values of Temkin constants A and B. In the adsorption of copper, nickel and zinc onto activated carbon produced from *Moringa oleifera* wood [31] the Temkin isotherm showed a higher correlation coefficient, which may be due to the linear dependence of the heat of adsorption on low or medium coverage. The repulsive force probably occurs between the different adsorbate species or for intrinsic surface heterogeneity may be associated with the linearity.

Table 4 details some of the results for the biosorption studies using *Moringa oleifera* which have been reported in the literature from 2006 onwards. From this table it is clear that *Moringa oleifera* shows versatility, removing a variety metals under favorable conditions and is among the most promising metal biosorbents.

Comparing the Langmuir and Freundlich models, *M. oleifera* seeds demonstrated a good removal capacity for Co(II), Cu(II), Pb(II), Cd(II) and Ag(I), as compared to reports related to other parts of the plant (Table 3). The variations in the removal percentage for metal ions can be explained by the different ionic radii of chemical species. In general, for the single metal solutions, ions with larger ionic radii are preferentially adsorbed. Among the metals tested, Pb(II) has the largest ionic radius and hence shows the highest adsorption percentage, whereas Co(II) presents the lowest level of adsorption [55].

Kinetics models are important in evaluating the basic qualities of an adsorbent as well as the time required for the removal of particular metals, the effectiveness of the adsorbent and the identification of the types of mechanisms involved in an adsorption system [56-58]. In order to investigate the mechanism of biosorption and its potential rate-controlling steps, which include the mass transfer and chemical reaction processes, kinetics models are exploited to test experimental data obtained in kinetics studies. These usually show an initial period of rapid metal adsorption with a subsequent decreased until reaching equilibrium of the system. This occurs due to the rapid adsorption of metallic ions by the surface of the adsorbent followed by a step of slow diffusion of ions from the surface film to the adsorption sites in the micropores which are less accessible [59].

In practice, the kinetic studies are carried out in batch experiments, typically varying the adsorbate concentration, the adsorbent mass, the agitation time and the temperature, as well as the type of adsorbent and adsorbate. Subsequently, the data are processed and used in the linear regression to determine the kinetics model which provides the best fit. However, for the validity of the order of the adsorption process two criteria should be evaluated, the first based on the regression coefficient (R^2) and the second on the calculated q_e values, which must approach the experimental q_e [60]. The main models used to evaluate the kinetics model profile are pseudo-first-order and pseudo-second order. However, other models are also are applied, such as Bangham's model and the Weber and Morris sorption kinetic model.

Table 4: Langmuir and Freundlich isotherm parameters for *Moringa oleifera*

Heavy metal	Langmuir model			Freundlich model			Ref.
	Qm (mg g⁻¹)	KL (L mg⁻¹)	R2	Kf (mg g⁻¹) (L mᵍ⁽¹/ⁿ⁾)	N (g L⁻¹)	R2	
Cd (II)	171.37	0.037 0.029	> 0.99	-	-	-	[32]
Cu(II)	167.90	0.023	> 0.99				
Ni(II)	163.88		> 0.99				
Ni(II)	30.38	0.31	0.9994	-	-	-	[35]
Cu(II)	11.534	0.2166	0.9979	3.8563	2.9214	0.9976	[31]
Zn(II)	17.668	0.1430	0.9528	3.7708	2.2528	0.9996	
Ni(II)	19.084	0.6165	0.9973	-	-	-	
Pb(II)	209.54	0.038	> 0.99	-	-	-	[34]
Ni(II)	29.6	-	0.9913	-	-	-	[44]
Ag(II)	23.13	0.1586	0.9935	-	-	-	[36]
Zn(II)	52.08	0.150	0.9994	50.35	-	0.9953	[16]
Cd(II)	1.06	0.51	0.94	-	-	-	[15]
Cr(III)	1.01	0.40	0.96				
Ni(II)	0.94	0.34	0.96				
As (III)	1.59	0.04	0.96				[43]
As (V)	2.16	0.09	0.98				
Pb(II)	-	-	0.9981	-	-	-	[3]
Pb (II)	-	-	-	8.6	2.8	0.9981	[19]
Cd(II)	-	-	-	3.04	1.37	-	[14]

The pseudo-first order equation, also known as the Lagergren equation, is expressed as follows [16,61]:

$$\log \left(q_e - q_t\right) = \log q_e - \frac{k_1}{2.303} t$$

(5)

where q_t and q_e (mg g^{-1}) are the amount of metal ions adsorbed per unit weight of the adsorbent at time t and equilibrium, respectively; and k_1 (min^{-1}) is the pseudo-first order rate constant of the sorption process and t (min) is the mixing time [60]. Table 5 presents the data of calculated q_e, pseudo-first order rate (k_1) and correlation coefficient (R^2). This kinetics model is based on the assumption that the adsorption rate is proportional to the number of free sites available, occurring exclusively onto one site per ion [34, 62].

In most studies discrepancies occurred between the value of q_e calculated by the pseudo-first order model and the experimental q_e, as shown in Table 5, highlighting the inability of this model to describe the kinetics of the metal ion adsorption processes. In general, calculated q_e values are smaller than the experimental q_e, which may occur because of a time lag, probably due to the presence of the boundary layer or external resistance at the beginning of the sorption process [63]. Considering the papers detailed in Table 5, only in [43] and [14] used only the pseudo-first order kinetics model to examine the data obtained, even though in the latter case the correlation values obtained were relatively low. In [43] noted no change in the adsorption rate constant when varying the concentrations of As(III) and As(V) and therefore this model could describe the adsorption process. In [14] used this model to compare the adsorption rate constants of ternary metal ions and single metal ions and noted that these constants were lower for ternary metal ions. Their explanation for this was that metal ions compete for vacant sites and uptake by binding sites within the shortest possible time.

The pseudo-second-order kinetics model is also based on the assumption that the sorption rate is controlled by a chemical sorption mechanism involving electron sharing or electron transfer between the adsorbent and adsorbate [64]. It can be expressed as:

$$\frac{t}{q_t} = \frac{1}{k_2 q_e^2} + \frac{t}{q_e}$$

(6)

where q_t and q_e (mg g^{-1}) are the amount of metal ions adsorbed per unit weight of the adsorbent at time t and equilibrium, respectively; and k_2 (g mg^{-1} min^{-1}) is the pseudo-second order rate constant of the sorption process and t (min) is the mixing time.

Table 5 presents the data of calculated q_e, pseudo-second order rate (k_2) and correlation coefficient (R^2). For most of the pseudo-second order kinetics models the calculated q_e values approach the experimental q_e values and the correlation coefficients are close to 1, indicating a good ability of this model to describe the kinetics of the metal ion adsorption process. This observation indicates that the rate-limiting steps in the biosorption of metallic ions are chemisorption involving valence forces through the sharing or exchange of electrons between the sorbent and the sorbate, complexation, coordination and/or chelation, in which mass transfer in the solution was not involved.

Considering that neither the pseudo-first-order nor the pseudo-second-order model can identify the diffusion mechanism, other kinetic models are needed to study this process, such as Bangham's model and the Weber and Morris sorption kinetics model [65]. The latter model is also known as the intra-particle diffusion model, this process in many cases being the rate-limiting step, which can be determined through the following equation:

$$q_t = k_{id} t^{1/2} + c_{id}$$

(7)

where q_t (mg g^{-1}) is the amount of metal ions adsorbed per unit weight of the adsorbent at time t, c_{id}(mg g^{-1}) is a constant of Weber and Morris, and k_{id} (mg g^{-1} min$^{-1/2}$) is the intra-particle diffusion rate constant and t (min) is the mixing time [66]. The value of the intercept gives an idea of the thickness of the boundary layer, i.e., the larger the intercept the greater the boundary layer effect will be. When there is a complete fit of the model the value of cid should be zero, and the deviation of this constant is due to differences in the mass transfer rate during the initial and final stages of adsorption. This is indicative that there is some degree of boundary layer control and shows that the intra-particle diffusion is not the only rate-limiting step, and thus several processes operating simultaneously may control the adsorption [34].

According to this model, if the plot of q_t versus $t^{1/2}$ gives a straight line, then the sorption process is controlled by intra-particle diffusion, while if the data exhibit multi-linear plots then two or more steps influence the adsorption process [67]. In two studies performed in [32, 19], multi-linear plots were observed with three distinct steps involved in the biosorption, the initial region of the curve relative relating to the adsorption on the external surface. The second region corresponds to the gradual uptake, where the intra-particle diffusion is the rate-limiting step. The final plateau region indicates the equilibrium uptake.

In reference [3] compared different types of carbon through the kid values and observed that the effect of intra-particle diffusion may be significantly increased by chemically modifying the adsorbents. Although none of the data collected in the studies detailed in Table 5 were well-described by the kinetics model proposed by Weber and Moris, the intraparticle diffusion may not be the only rate-limiting step in these studies.

Bangham's model evaluates whether pore diffusion is the only rate-controlling step in the adsorption process [65] and can be represented by the following equation:

$$\log \left(\frac{c_0}{c_0 - q_t m} \right) = \log \left(\frac{k_0 m}{2.303 \, V} \right) + \sigma$$

(8)

where q_t (mg g^{-1}) is the amount of metal ions adsorbed per unit weight of the adsorbent at time t; co(mg L^{-1}) is the initial metal ion concentration in liquid phase; m (g L^{-1}) is adsorbent concentration at time t (min); V (L) solution volume, t (min) is the mixing time and k_0 (L g^{-1}) and (< 1) are constants of Bangham's model. Of the studies published, only in [3] used Bangham's kinetic model to compare the rate constants for the adsorption of Pb(II) onto different types of functionalized carbon prepared from the seed husks and pods of *M. oleifera* and thereby assess the efficiency of the functionalization of this material.

The temperature is reportedly an important parameter for the adsorption of metal ions. An increase or decrease in temperature can cause a change in the amount of metal removed or adsorbed by the adsorbent. A change in temperature causes a change in the thermodynamic parameters of free energy ($\Delta G°$), enthalpy ($\Delta H°$) and entropy ($\Delta S°$). These parameters are important to understand the adsorption mechanism [68]. For a given temperature, a phenomenon is considered to be spontaneous if the $\Delta G°$ has a negative value. Moreover, if $\Delta H°$ is positive the process is endothermic and if it is negative the process is exothermic [69]. Negative values of $\Delta S°$ show a decreased randomness or increased order at the metal-biomass interface. The positive value showed a change in the biomass structure during the sorption process, causing an increase in the disorder of the system [68]. The parameters $\Delta G°$ (kJ mol^{-1}), $\Delta H°$ (kJ mol^{-1}) and $\Delta S°$ (J mol^{-1} K^{-1}) can be evaluated from the following equations [70].

$$\Delta G° = -RT \ln K_c$$

(9)

$$\ln K_c = -\frac{\Delta H°}{RT} + \frac{\Delta S°}{R}$$

(10)

where R(8.314J mol^{-1} K^{-1}) is the gas constant, T(K) the absolute temperature and K_c(mL g^{-1}) the standard thermodynamic equilibrium constant defined by qe/Ce. $\Delta H°$ and $\Delta S°$ can be determined from the slope and the intercept of the linear plot of Ln K_c versus $1/T$.

The studies performed on *Moringa oleifera* using chemically-modified leaves for the adsorption of Pb(II) [34], bark for Ni(II) [35] and leaves for Cd(II), Cu(II) and Ni(II) [32] showed the endothermic nature and spontaneity of the adsorption process. The positive values of $\Delta S°$ suggest an increase in randomness at the solid/liquid interface with some structural changes in the sorbate.

FINAL CONSIDERATIONS

Although the biosorption of heavy metals from aqueous solutions is a relatively new process that has proven very promising in the removal of contaminants from aqueous effluents, offering significant advantages like the low-cost, availability, profitability, easy of operation and efficiency. Other technologies have also been very attractive ensuring an appropriate process to treat industrial waste effluents [71-77]. However, biosorption is becoming a potential alternative to the existing technologies for the removal and/or recovery of toxic metals from wastewater. The major advantages of biosorption technology are its effectiveness in reducing the concentration of heavy metal ions to very low levels and the use of inexpensive biosorbent materials.

CONCLUSIONS

The studies described herein indicate that *Moringa oleifera* seeds are an alternative sorbent for metal ion removal from contaminated waters. This can be found in most papers which report 60 to 90% removal of metals (Cd(II), Cu(II), Ni(II), Pb(II), As(III), As(V), Cr(III) and Zn(II)). In these cases, not only the seeds were used, but also leaves, bark and pods showing the great versatility of this plant. The results show that even with the high heterogeneity of the matrix confirmed through characterization techniques there is a great potential for the application of these seeds in effluent treatment without component separation, which makes the process economically and technically attractive.

Table 5: Kinetics parameters for metal biosorption using Moringa oleifera

Metal	c_o (mg L^{-1})	qe, exp (mg g^{-1})	Pseudo-first-order			Pseudo-second-order				Ref.
			Qe (mg g^{-1})	k1 (min^{-1})	R2	qe (mg g^{-1})	k2 (g mg^{-1}min^{-1})	h_0 (mg g^{-1}min^{-1})	R2	
Zn (II)	50[a]	45.00	-	-	-	46.94	4.51 10-4	-	0.997	[16]
	50[b]	45.76	-	-	--	47.16	2.85 10-4	-	0.999	
	5[c]	42.80	-	-	-	43.47	2.04 10-5	-	0.997	
Pb(II)	30[d]	-	-	-	-	24.57	0.0085	5.131	0.999	[3]
	30[e]	-	-	-	-	27.70	0.0052	3.989	0.998	
	30[f]	-	-	-	-	28.49	0.0060	4.868	0.998	
	30[g]	-	-	-	-	29.08	0.0062	5.243	0.999	
	3°h	-	-	-	-	29.46	0.0087	7.550	0.999	
As (III)	25	-	-	0.047	-	-	-	-	-	[43]
As (V)	50	-	-	0.049	-	-	-	-	-	
	25	-	-	0.063	-	-	-	-	-	
	50	-	-	0.065	-	-	-	-	-	
Cd(II)	25	1.06	-	0.51	-	-	-	-	-	[14]
Cr(III)	25	1.01	-	0.40	-	-	-	-	-	
Ni(II)	25	0.94	-	0.34	-	-	-	-	-	

Pb(II)	10	12.7343	-	-	13.26	15.35	0.20	0.9974	[34]
	2.5	19.8988	-	-	20.64	10.08	0.21	0.997	
	40	23.9233	-	-	25.01	9.3	0.23	0.9995	
Pb(II)	10.4	8.7	-	-	8.8	27.8	2.5	0.9999	[19]
	30.1	10.2	-	-	10.3	18.2	1.9	0.9999	
	50.4	12.5	-	-	12.53	12.6	1.6	0.9998	
Ni(II)	10	9.7	-	-	10.29	1.91	2.03	0.9971	[35]
	2.5	6.74	-	-	7.14	2.70	1.38	0.9964	
	50	3.27	-	-	3.43	12.74	1.05	0.996	
Cu(II)	30	8.3406	-	-	8.3264	0.0848	-	0.9998	[31]
Zn(II)	30	13.2537	-	-	13.2450	0.2457	-	1	
Ni(II)	30	9.5847	-	-	9.6154	0.0957	-	0.9999	
Cd (II)	10	13.54	-	-	10.99	1.39	2.73	0.9951	[32]
	25	13.80	-	-	10.50	1.46	3.09	0.9969	
	40	20.86	-	-	15.24	1.22	5.61	0.9981	
Cu(II)	10	11.92	-	-	1.03	3.4	5.58	0.9992	
	25	13.55	-	-	12.45	1.66	3.37	0.9958	
	40	16.01	-	-	12.86	1.56	4.31	0.9977	
Ni(II)	10	10.24	-	-	10.24	1.51	1.70	0.9951	
	25	12.49	-	-	12.49	1.31	2.29	0.9952	
	40	14.07	-	-	14.07	1.52	3.27	0.9967	

Biosorption is the most economical and eco-friendly method for the removal of heavy metals from domestic as well as industrial wastewater and it is particularly important to promote the development of biosorption for industrial processes. Notable advantages are: (a) low cost of the biosorbent, (b) high efficiency for metal removal at low concentration, (c) potential for biosorbent regeneration and metal valorization, (d) high sorption and desorption rates, (e) limited generation of secondary residues, and (f) relatively environmentally-friendly life cycle of the material (easy to eliminate compared to conventional resins, for example).

However, after the metal removal from aqueous solutions by the biomass, the recovery of the metal is an important issue. This can be achieved through a metal desorption process, aimed at weakening the metal-biomass linkage. Thus, studies to evaluate the reversibility of the adsorption reactions involved in the biosorption of heavy metals are of great importance. The problems associated with the disposal of exhausted adsorbent can be solved either by its activation or incineration or its disposal after proper treatment. For biosorption and desorption processes, another important aspect is the biosorbent reuse in successive biosorption-desorption cycles, the viability of which is determined by the cost-benefit relationship between the loss in biosorption capacity during the desorption steps and the operational yield in the metal recovery. Thus, further studies need to focus on the development of new clean environmentally-acceptable technologies.

ACKNOWLEDGEMENTS

The authors are grateful for financial support from the government agencies Conselho Nacional de Desenvolvimento Científico e Tecnológico (CNPq), Fundação de Amparo à Pesquisa do Estado de Minas Gerais (FAPEMIG), Fundação de Amparo à Pesquisa do Estado de Goiás (FAPEG) and Coordenação de Aperfeiçoamento de Pessoal de Nível Superior (CAPES).

REFERENCES

1.   Miliarum: Ingenieria Civil y Medio Ambiente. Cursos Edafología on line: Contaminación por metales pesados. http://www. miliarium.com/Marcos/SuelosContaminados.asp (accessed 27 September 2012).

2.   Duffus JH. "Heavy metals" - A meaningless term? (IUPAC Techinical Report). Pure and Applied Chemistry 2002; 74(5) 793-807.

3.   Nadeem M, Mahmooda A, Shahid SA, Shah SS, Khalid AM, McKaye G. Sorption of lead from aqueous solution by chemically modified carbon adsorbents. Journal of Hazardous Materials 2006; 138(3) 604-13.

4.   Obuseng V, Nareetsile F, Kwaambwa HM. A study of the removal of heavy metals from aqueous solutions by Moringa oleifera seeds and amine-based ligand 1,4-bis[N,N-bis(2-picoyl)amino] butane. Analytica Chimica Acta 2012; 730 87-92.

5.   Zevenhoven R, Kilpinen P. Control of pollutants in flue gases and fuel gases. Finland: Espoo; 2001.

6.   EPA: Environmental Protection Agency. National Primary Drinking Water Regulations. http://www.biovir.com/Images/pdf054.pdf (accessed 27 September 2012).

7.   Babel S, Kurniawan TA. Cr(VI) removal from synthetic wastewater using coconut shell charcoal and commercial activated carbon modified with oxidizing agents and/or chitosan. Chemosphere 2004; 54(7) 951-67.

8.   Bai RS, Abraham TE. Low-cost supports used to immobilize fungi and reliable technique for removal hexavalent chromium in wastewater. Bioresource Technology 2003; 87(17) 17-26.

9.   Petroni SLG, Pires MAF, Munita CS. Adsorção de zinco e cádmio em colunas de turfa. Química Nova 2000; 23(4) 477-81.

10.   Volesky B. Detoxification of metal-bearing effluents: biosorption for the next century. Hydrometallurgy 2001; 59(2) 203-16.

11.   Naja G, Murphy V, Volesky B. Biosorption, Metals - Encyclopedia of Industrial Biotechnology. United Kingdom: Wiley; 2010.

12.   Melcakova I, Horvathova H. Study of biomass of reynoutria japonica as a novel biosorbent for removal of metals from

aqueous solutions. Czech Republic: GeoScience Engineering; 2010. http://gse.vsb.cz/2010/LVI-2010-1-55-70.pdf (accessed 27 September 2012).

13. Dermirbas, A, Heavy metal adsorption onto agro-based waste materials: A review. Journal of Hazardous Materials 2008; 157 220-29

14. Sharma P, Kumari P, Srivastava MM, Srivastava S. Removal of cadmium from aqueous system by shelled Moringa oleifera Lam seed powder. Bioresource Technology 2006; 97 299-305.

15. Sharma P, Kumari P, Srivastava MM, Srivastava S. Ternary biosorption studies of Cd(II), Cr(III) and Ni(II) on shelled Moringa oleifera seeds. Bioresource Technology 2007; 98 474-77.

16. Bhatti HN, Mumtaz B, Hanif MA, Nadeem R. Removal of Zn(II) ions from aqueous solution using Moringa oleifera Lam. (horseradish tree) biomass. Process Biochemistry 2007; 42 547-53.

17. Keer WE, Silva AR. "Moringa: uma nova hortaliça para o Brasil". Uberlândia: UFU/ DIRIU, 1999.

18. Almeida ILS. Avaliação da capacidade de adsorção da torta da Moringa oleifera para BTEX em amostras aquosas. Masters dissertation. Federal University of Uberlândia; 2010.

19. Reddy DHK, Seshaiah K, Reddy AVR, Rao MM, Wang MC. Biosorption of Pb2+ from aqueous solutions by Moringa oleifera bark: equilibrium and kinetic studies. Journal of Hazardous Materials 2010; 174 831-838.

20. Nautiyal BP, Venkataraman KG. Moringa (drumstick) - an ideal tree for social forestry: growing conditions and uses - part 1. Myforest 1987; 23(1) 53-8.

21. Ndabigengesere A, Narasiah KS, Talbot BG. Active agents and mechanism of coagulation of turbid waters using Moringa oleifera. Water Research 1995; 29(2) 703-10.

22. Jahn SAA. Proper use Moringa oleifera for food and water purification - Selection of clones and growing of annual short-stem. Pflanzenzucht 1989; 4 22-5.

23. Oliveira JTA, Silveira SB, Vasconcelos IKM, Cavada BS, Moreira RA. Compositional and nutritional attributes of seeds from the

multiple purpose tree Moringa oleifera Lamarck. Journal of the Science of Food and Agriculture 1999; 79 815-20.

24. Olsen A. Low technology water purification by bentonite clay and Moringa oleifera seed flocculation as performed in Sudanese village: effects on Schistoma mansoni cercariae. Water Research 1987; 21(5) 517-22.

25. Madsen M, Schumundt J, Omer EFE. Effect of watter coagulation by seeds of Moringa oleifera on bacterial concentration. Journalof tropical Medicine and Hygiene 1987; 90(3)101-9.

26. Bennett RN, Mellon FA, Foidl N, Pratt JH, Dupont MS, Perkins L, Kroon A. Profiling Glucosinolates and Phenolics in Vegetative and Reproductive Tissues of the Multi-Purpose Trees Moringa oleifera L. (Horseradish Tree) and Moringa stenopetala L. Journal of Agricultural and Food Chemistry 2003; 51(12) 3546-53.

27. Moreira DR. Desenvolvimento de adsorventes naturais para tratamento de efluentes de galvanoplastia. Masters dissertation. Pontifical Catholic University of Rio Grande do Sul; Porto Alegre; 2010.

28. Davis TA, Mucci A, Volesky B. A review of the biochemistry of heavy metal biosorption by brown algae. Water Research 2003; 37(18) 4311-30.

29. Schneider IAH, Rubio J. Plantas Aquáticas: Adsorventes Naturais para a Melhoria da Qualidade das Águas, 2003. XIX Prêmio Jovem Cientista - 2003 – Água: Fonte de Vida, Porto Alegre, Brazil, School of Engineering, Federal University of Rio Grande do Sul, 2003.

30. Pagnanelli F, Mainelli S, Veglio F, Toro L. Heavy metal removal by olive pomace: biosorbent characterization and equilibrium modeling. Chemical Engineering Science 2003; 58 4709-17.

31. Kalavathy MH, Miranda LR. Moringa oleifera—A solid phase extractant for the removal of copper, nickel and zinc from aqueous solutions. Chemical Engineering Journal 2012; 158 188–99.

32. Reddy DHK, Seshaiaha K, Reddyb AVR, Leec SM. Optimization of Cd(II), Cu(II) and Ni(II) biosorption by chemically modified Moringa oleifera leaves powder. Carbohydrate Polymers 2012; 88 1077–86.

33. Brito ES, Damasceno LF, Gallão MI. Avaliação química e estrutural da semente de moringa. Revista Ciência Agronômica 2006; 37 106-9.

34. Reddy DHK, Harinatha Y, Seshaiaha K, Reddy AVR. Biosorption of Pb(II) from aqueous solutions using chemically modified Moringa oleifera tree leaves. Chemical Engineering Journal 2010; 162 626–34.

35. Reddy DHK, Ramana DKV, Seshaiah K, Reddy AVR. Biosorption of Ni(II) from aqueous phase by Moringa oleifera bark, a low cost biosorbent. Desalination 2011; 268 150–57.

36. Araújo CST, Melo EI, Alves VN, Coelho NMM. Moringa oleifera Lam. seeds as a natural solid adsorbent for removal of AgI in aqueous solutions. Journal of the Brazilian Chemical Society 2010; 21 1727-32.

37. Fifield FW, Kealey D. Principles and Practice of Analytical Chemistry. Oxford: Blackwell Science; 2000.

38. Anwar F, Rashid U. Physico-chemical characteristics of Moringa oleifera seeds and seeds oil from a wild provenance of Pakistan. Pakistan Journal of Botany 2007; 39(5) 1443-53.

39. Pereira GM, Arruda MAZ. Trends in preconcentration procedures for metal determination using atomic spectrometry techniques. Microchimica Acta 2003; 141 115-31.

40. Senthilkumaar S, Bharathi S, Nithyanandhi D, Subburam V. Biosorption of toxic heavy metals from aqueous solutions. Bioresource Technology 2000; 75 163-5.

41. Alves VN, Mosquetta R, Coelho NMM, Bianchin JN, Roux KCP, Martendal E, Carasek E. Determination of cadmium in alcohol fuel using Moringa oleifera seeds as a biosorbent in an on-line system coupled to FAAS. Talanta 2010; 80(3) 1133-38.

42. Marshall WE, Johns MM. Agricultural by-products as metal adsorbents: sorption properties and resistance to mechanical abrasion. Journal of Chemical Technology 1996; 66(2) 192-8.

43. Kumari P, Sharma P, Srivastava S, Srivastava MM. Biosorption studies on shelled Moringa oleifera Lamarck seed powder: removal and recovery of arsenic from aqueous system. International Journal Mineral Process 2006; 78 131-9.

44.   Marques TL, Alves VN, Coelho LM, Coelho NMM. Removal of Ni(II) from aqueous solution using Moringa oleifera seeds as a bioadsorbent. Water Science & Technology 2012; 68 1435-40.

45.   Shen D, Fan J, Zhou W, Gao B, Yue Q, Kang Q. Adsorption kinetics and isotherm of anionic dyes onto organo-bentonite from single and multisolute systems. Journal of Hazardous Materials 2009; 172 99-107.

46.   Abdullah MA, Chiang L, Nadeem M. Comparative evaluation of adsorption kinetics and isotherms of a natural product removal by Amberlite polymeric adsorbents. Chemical Engineering Journal 2009; 146 370-6.

47.   Ma X, Li L, Yang L, Su C, Wang K, Yuan S, Zhou J. Adsorption of heavy metals ions using hierarchical CaCO3-maltose meso/macroporous hybrid materials: Adsorption isotherms and kinetic studies. Journal of Hazardous Materials 2012; 209-210 467-477.

48.   Tofighy MA, Mohammadi T. Adsorption of divalent heavy metal ions from water using carbon nanotube sheets. Journal of Hazardous Materials 2011; 185 140-7.

49.   Olu-owolabi BI, Oputu OU, Adebowale KO, Ogunsolu O, Olujimi OO. Biosorption of Cd2+ and Pb2+ ions onto mango stone and cocoa pod waste: Kinetic and equilibrium studies. Scientific Research and Essays 2012; 7(15) 1614-29.

50.   Kumar PS, Ramalingam S, Senthamarai C, Niranjanaa M, Vijayalakshmi P, Sivanesan S. Adsorption of dye from aqueous solution by cashew nut Shell: Studies on equilibrium isotherm, kinetics and thermodynamics of interactions. Desalination 2010; 261 52-60.

51.   Mahmoud DK, Salleh MAM, Karim WAA. Langmuir model application on solid-liquid adsorption using agricultural wastes: Environmental application review. Journal of Purity, Utility Reaction and Environment 2012; 1(4) 170-99.

52.   Nwabanne JT, Igbokwe PK. Mechanism of cooper (II) removal from aqueous solution using activated carbon prepared from different agricultural materials. International Journal of Multidisciplinary Sciences and Engineering 2012; 3(7) 46-52.

53.   Kumar PS, Ramalingam S, Kirupha SD, Murugesan A, Vidhyadevi T, Sivanesan S. Adsorption behavior of nickel (II) onto cashew nut

Shell: equilibrium. Thermodynamics, kinetics, mechanism and process design. Chemical Engineering Journal 2011; 167 122-31.

54. Kumar PS, Ramalingan S, Sathyaselvabala V, Kirupha SD, Murugesan A, Sivanesan S. Removal of cadmium(II) from aqueous solution by agricultural waste cashew nut shell. Korean Journal of Chemical Engineering 2012; 29(6) 756-68.

55. Matos GD, Arruda MAZ. Online preconcentration/determination of cadmium using grape bagasse in a flow system coupled to thermospray flame furnace atomic absorption spectrometry. Spectroscopy Letters 2006; 39 1-14.

56. Febrianto J., Kosasih A. N., Sunarso J., Ju. Y., Indraswati N., Ismadji S. Equilibrium and kinetic studies in adsorption of heavy metals using biosorbent: A summary of recent studies. Journal of Hazardous Materials 2009; 162, 616-45.

57. Kumar KV, Sivanesan S. Pseudo-second-order kinetic models for safranin onto rice husk: comparison of linear and non-linear regression analysis. Process Biochemistry 2006; 41 1198–1202.

58. Savic JZ, Vasic VM. Thermodynamics and kinetics of 1,8-dihydroxy-2-(imidazol-5-yl-azo)-naphthalene-3,6-disulphonic acid immobilization on Dowex resin, Colloids and Surfaces A. Physicochemical and Engineering Aspects 2006; 278 197–203.

59. Zhao G, Wu X, Tan X, Wang X. Sorption of Heavy Metal Ions from Aqueous Solutions: A Review. The Open Colloid Science Journal 2011; 4 19–31.

60. Li X, Zheng W, Wang D, Yang Q, Cao J, Yue X, Shen T, Zeng G. Removal of Pb (II) from aqueous solutions by adsorption onto modified areca waste: Kinetic and thermodynamic studies. Desalination 2010; 258 148–53.

61. Bark N, Abdennouri M, Boussaoud A, Makhfouk MEL. Biosorption characteristics of Cadmium (II) onto Scolymus hispanicus L. as low-cost natural biosorbent. Desalination 2010; 258(1-3) 66–71.

62. Lalhruaitluanga H, Jayaram K, Prasad MNV, Kumar KK. Lead(II) adsorption from aqueous solutions by raw and activated charcoals of Melocanna baccifera Roxburgh (bamboo)-A comparative study. Journal of Hazardous Materials 2010; 75 311–8.

63. Vijayaraghavan K, Palanivelu K, Velan M. Biosorption of copper(II) and cobalt(II) from aqueous solutions by crab shell particles, Bioresource Technology 2006; 97 1411–19.

64. Ding Y, Jing D, Gong H, Zhou L, Yang X. Biosorption of aquatic cadmium(II) by unmodified rice straw. Bioresource Technology 2012; 114 20–5.

65. Yaneva Z, Koumanova B, Georgieva N. Study of the Mechanism of nitrophenols sorptions on expanded perlite – Equilibrium and Kinetics Modeling. Macedonian Journal of Chemistry and Chemical Engineering 2012; 31(1) 101–14.

66. Baidas S, Gao B, Meng X. Perchlorate removal by quaternary amine modified reed. Journal of Hazardous Materials 2011; 189 54–61.

67. Bilgili MS. Adsorption of 4-chlorophenol from aqueous solutions by xad-4 resin: isotherm, kinetic, and thermodynamic analysis. Journal of Hazardous Materials 2006; 137 157–64.

68. Farooq U, Kozinski JA, Khan MA, Athar M. Biosorption of heavy metals ions using wheat based biosorbents - A review of the recent literature. Bioresource Technology 2010; 101 5043-53.

69. Gueu S, Yao B, Adouby K, Ado G. Kinetics and thermodynamics study of lead adsorption on to activated carbons from coconut and seed hull of the palm tree. International Journal Science Technology 2007; 4(1) 11-17.

70. Zhu HY, Fu YQ, Jiang R, Jiang JH, Xiao L, Zeng GM, Zhao SL, Wang Y. Adsorption removal of congo red on magnetic cellulose/ Fe3O4/activated carbon composite: Equilibrium, kinetic and thermodynamic studies. Chemical Engineering Journal 2011; 173 494-502.

71. Bhat V, Rao P, Patil Y. Development of an integrated model to recover precious metals from electronic scrap - A novel strategy for e-waste management. Procedia - Social and Behavioral Sciences 2012; 37: 397-406.

72. Patil YB. Development of an innovative low-cost industrial waste treatment technology for resource conservation - A case study with gold-cyanide emanated from SMEs. Procedia- Social and Behavioral Sciences 2012; 37: 379-388.

73. Gaddi SS, Patil YB. Screening of some low-cost waste biomaterials for the sorption of silver-cyanide [Ag(CN)2-] from aqueous solutions. International Journal of Chemical Sciences 2011; 9: 1063-1072.

74. Patil YB, Paknikar KM. Biological detoxification of nickel-cyanide from industrial effluents. Process Metallurgy 2001; 11: 391-400.

75. Patil YB, Paknikar KM. Development of a process for biodetoxification of metal cyanides from wastewaters. Process Biochemistry 2000; 35: 1139-1151.

76. Patil YB, Paknikar KM. Biodetoxification of silver-cyanide from electroplating industry wastewater. Letters in Applied Microbiology 2000; 30: 33-37.

77. Patil YB, Paknikar KM. Removal and recovery of metal cyanides using a combination of biosorption and biodegradation processes. Biotechnology Letters 1999; 21: 913-919.

Reclamation of Degraded Landscapes Due to Opencast Mining

Nazan Kuter

¹Cankiri Karatekin University, Faculty of Forestry, Department of Landscape Architecture, and Cankiri, Turkey

INTRODUCTION

Even though it is regarded as a crucial economic activity worldwide, mining has a significant negative impact on environment. Due to its nature, especially opencast mining inevitably leads to serious degradation on ecological and aesthetic values of the landscape.

Topography and drainage, air, soil and water quality, vegetation including forest ecosystems, noise levels and ground vibrations, human health and habitation can be listed as the typical parameters that are mainly affected by opencast mining activities. When the extraction of reserve is over, the altered landscape has to be reclaimed in order to relieve the damaging effects of opencast mining and restore the landscape and its immediate surroundings.

On the other hand, reclamation of post-mining landscapes is a very challenging task since there is no unique reclamation planning scheme for such landscapes, and it highly depends on the site-specific characteristics. Therefore, successful and sustainable reclamation requires interdisciplinary approach leading to an integrated and effective proposal to restore ecological, hydrological, aesthetic, recreational and other functions of the post-mining landscape. Different methods and approaches for the reclamation of opencast mine sites have been proposed by several disciplines such as landscape architecture, environmental and mining engineering, forestry, archeology and social sciences.

The main motivation of this chapter is to emphasize both the importance of reclamation studies and the fact that natural and cultural characteristics of the post-mining landscapes have to be considered within different point of views by various disciplines simultaneously in order to obtain the most suitable landscape use planning for such areas.

The remainder of this chapter is organized as follows. The next section gives basic overview of the effects of opencast mining activities on both environment and human health. In Section 3, reclamation and rehabilitation are addressed within a broader perspective, including the definition of basic terminology; the aim, the importance and the necessity for the reclamation of opencast mine sites; methods and techniques; evaluation of the success of reclamation; interdisciplinary dimension of the issue; several case studies; and legislative matters as well. Finally, Section 4 concludes this chapter.

THE EFFECTS OF OPENCAST MINING ACTIVITIES ON ENVIRONMENT AND HUMAN HEALTH

Mining is important for local and global economy, but this operation mostly and inevitably leads to substantial environmental damage and due to these kinds of activities, original potential of landscape is extremely altered.

Especially in the case of opencast mining, where a mineral is fairly close to the surface in a massive or wide tabular body, or the mineral itself is part of the surface soil or rock, surface mining methods are often considered as more economical. The most common surface mining methods such as strip mining, open pit mining, opencast mining and quarrying start from the earth's surface and keep exposure to the surface during the extraction period. Disruption of the surface significantly affects the soil, fauna, flora and surface water, thereby influencing all types of land use. Additionally, if the operation goes further below the water table, it will affect the near-surface groundwater (Chamber of Mines of South Africa 2008).

Most surface mining methods are large scale, involving removal of massive volumes of material, including overburden, to extract the mineral deposit. Large amounts of waste can be produced in the process. Surface mining also can cause noise and disturbance, leave scars on the landscape and may pollute the air with dust (Bell and Donnelly 2006). Therefore, it is not only crucial to have a detailed understanding of the pre-mining environment, but also important to apprehend the utilized mining method in order to plan a meaningful surface rehabilitation, wherever possible (Chamber of Mines of South Africa 2008). The process of removing, storing and subsequently replacing the soil during the mining activity lead to potential problems in relation to subsequent restoration. In this respect, a major distinction should be drawn between those sites where, for operational reasons, soil has to be stored for a period of years while the mining progresses, and those, usually larger, sites where a progressive system of restoration can be practiced (Rimmer and Younger 1997).

The negative impacts of surface mining on environment can be listed as the following (Kavourides et al. 2002):

- occupation of large farming areas needed for excavation and dumping operations,
- alteration of land morphology,
- disturbance of native fauna and flora,
- modification of surface and ground water balance,
- resettlement of residential areas, roads and railways,
- release of air, liquid and solid pollutants and noise pollution.

Water resources and the quality of air are seriously modified by surface mining operations. One problem introduced during surface mining operations is groundwater, which contains dissolved salts derived from the rock that it has been in contact with, and it is characterized according to the concentrations and proportions of combinations of ions that it contains. Impacts of surface mining are often large and unpredicted such as a former zinc-copper mine polluting the environment due to cadmium leachates or a former gold-copper ore causing arsenic pollution of surface waters (Sengupta 1993; Sams and Beer 2000; Salonen et al. 2003; Bell and Donnelly 2006).

Pöykiö et al. (2002) have evaluated the impact of a chromium opencast mining complex on the ambient air environment at Kemi, Northern Finland. The total suspended particles and associated metal (Cr, Ni and Pb) concentrations in the air were determined in their study area.

Soil destruction is one of the most crucial environmental impacts of opencast mining activities. In the course of removing the desired mineral material, original soil become lost, or buried by wastes. When mining is going and has gone on, particularly top soil must be conserved because it is an essential source of seed and nutrients, and should be preserved for use in reclamation. According to Mummey et al. (2002), disturbance of soil ecosystems that disrupts normal functioning or alters the composition of soil microbial communities is potentially destructive for both short and long term ecological stability.

Surface mining speeds up erosion and sedimentation and short duration, high intensity storms can be a violent force moving thousands of tons of soil. Physical characteristics of the overburden, degree and length of slope, climate, amount and rate of rainfall, type and percentage of vegetative ground cover affect the vulnerability of strip mined land erosion (Sengupta 1993). According to the Kleeberg et al. (2008); soil erosion is frequently related to high rates of particulate phosphorus (P) transfer from land to water bodies. Providing a long term source of P for aquatic biota, and accelerating freshwater eutrophication, information on P sources is important for good environmental management. In their study, a year-long monitoring, and ten short rainfall simulations on plot scale, at ridges and rills and a combination of them, revealed high erosion from bare lignite mining dumps at Schlabendorf-North, Lusatia, Germany.

Another adverse impact of opencast mining on land is soil contamination with a range of potentially hazardous substances (both chemical and biological) which, if present at sufficiently high levels, may introduce potential problems related to public health and environment. For example, soils can contain high levels of heavy metals such as cadmium and lead, which can severely affect the local population (Kibble and Saunders 2001). So, identifying and dealing with contaminated land is important in order to support increased quality of life for communities and conservation of biodiversity (Kibblewhite 2001).

It has been recently declared by the *United States Environmental Protection Agency* (*US EPA*) that the imperfect management of wastes produced in the course of mining and reclamation works is detrimental to environmental and human health. The effect of wastes due to mining and processing activities on ecosystems can be observed in groundwater, surface water, and soil and the following points may put human health in danger (Passariello et al. 2002):

* inhalation of aerosols containing high levels of metals,
* percutaneous absorption following skin contact,
* use of contaminated water,
* consumption of food from contaminated areas.

As a result of the study of Coelho et al. (2007), it has been stated that irritating symptoms have been found in the eye mucous and respiratory system of people living near abandoned mine pits, and population in the Vila Real district, in the Northeast of Portugal have been exposed to higher level of lead and cadmium.

Razo et al. (2004) assessed the environmental impact of arsenic and heavy metal pollution of soil, sediment and surface water in the Villa de la Paz-Matehuala, San Luis Potosi in Mexico, and the results of soil samples reported high concentrations of chemicals hazardous to human health. In order to give a specific example, the maximum arsenic concentration in pluvial water storage ponds (265 $\mu g.L^{-1}$), near the main potential sources of pollution, exceed by 5 times the Mexican drinking water quality guideline (50 $\mu g.L^{-1}$).

It is a matter of necessity at this point to both emphasize and focus on the negative effects of opencast coal mining on ecosystems at a landscape level, which may not only be large scale, but also be intense.

Environmental impacts of opencast coal mining have been thoroughly investigated by many researchers and defined for the various stages of the coal fuel cycle. The "coal cycle" comprises five main activities: i) Exploration and extraction; ii) Preparation; iii) Handling and supply; iv) Conversion (where applicable); and v) Utilization, including waste disposal. The principal environmental impacts and concerns specific to exploration, extraction, and preparation phases are listed below (Buchanan and Brenkley 1994):

- Surface mines: siting; large-scale land use; overburden removal and disposal; disturbance of hydrology and run-off; acid mine drainage; visual intrusion; noise; blast vibration; fly rock; fugitive dust; transportation/traffic; high wall stability; restoration of soil fertility; recreating ecosystem diversity; recreating landscape; amenity value; historic resource preservation.

- Abandoned mines: methane migration; flooding; groundwater contamination; structural integrity; land rehabilitation.

The environmental impacts resulting from coal mining activities are mainly attributable to the exposure of decreased earth materials, especially such as coal, pyrite, siderite, and ankerite, and to the oxidizing power of the Earth's atmosphere. The consequences range from the spontaneous combustion of coal to the release of acidic waters from pyrite oxidation. If no extenuating measures are used, potentially many unpleasant environmental impacts result from surface coal mining area. A typology of the known impacts resulting from mine voids and wastes in coal mining districts has been developed, which recognizes many subcategories of impacts such as air pollution, ground deformation, water pollution and water resource depletion (Sengupta 1993; Younger 2004).

According to Ghose (2002), opencast coal mining causes much more environmental pollution especially air quality deterioration in respect of dust and gaseous pollutants. It creates air pollution problem in the mining premises and the surrounding locations. In the study, the sources of air pollution in Jharia Coalfield, Indiana were identified, and *Suspended Particulate Matter* (*SPM*) and *Respirable Particulate Matter* (*RPM*) concentrations were found to be very high in work zone as well as surrounding locations. The study emphasized that stringent air quality standards should be set for coal mining areas.

In Sokolov coal mining district in Czech Republic, total area of more than 6000 ha will have been disturbed around year 2036 at the end of mining activities. Spoil material overlying the coal layer was removed and deposited in heaps. The largest heaps formed by removal of spoil material are thousands of hectares in the area and reach elevations of more than 100 m above the original terrain (Frouz et al. 2006).

In the Lusatian mining district of eastern Germany, where 6% of the global lignite production occurred during 90s, this influence is of particular concern. Over the last hundred years 75,000 ha of land have been turned into dumps. The water balance of the whole region has been changed by groundwater pumping. Fifty percent of the dump area was not reclaimed by the year 1998. At many places recultivation efforts were impeded by extreme ecological site conditions mainly due to the high pyrite content of the spoil material (Hüttl 1998).

Xin-yi et al. (2009) investigated Yanma coal mining waste dump in China in their study. The surface layer soil around the mountain was gathered, and the heavy metal content and pH were measured out. The heavy metal (Pb, Zn, Cu, Cr, Cd) pollution situation of the soil was researched according to the distance of coal mining waste dump. As revealed out from the study, heavy metal polluted the soil in certain distance to the coal mining waste dump, and the content is in negative correlation with the distance to the coal mining waste dump.

Bell et al. (2001) studied the environmental degradation associated with the abandoned Middelburg Colliery in the Witbank Coalfield, South Africa. The chemical composition of spoil materials of the mine mainly consist of two principal oxides: silica and alumina; calcium, magnesium, iron, sodium, potassium, and titanium oxides are also present in small concentrations. Pyrite takes place in the shales and coal of the spoil heaps, and its contact with air gives a toxic nature to soil heaps, which is not in favor of healthy vegetation growth and plant life.

The chemistry of groundwater in contact with coal mine workings may change due to reactions with iron pyrite, which may result either from oxidation of pyritic materials increasing the acidity of the water, or from dissolution of soluble salts in the spoil, overburden, or increasing levels of dissolved solids in the water. The oxidation process requires that both air and water come in contact with pyritic materials, whereas air is not required for dissolution of soluble salts. As a result of these chemical changes, groundwater becomes highly ferruginous and often has a low pH value, and hence it is referred to as *acid mine drainage*, which can be toxic due to high values of sulphate and increased levels of heavy metals. Where this groundwater flows into surface watercourses, the latter may become grossly polluted, and it may also cause other problems such as faults and subsidence being reactivated or the displacement and emergence of mine gases into the environment. *Acid mine drainage* is a significant, unremedied environmental problem which deteriorates surface and ground water quality. Also, it is of value to notice that some of the closed mine sites under investigation still cause severe environmental degradation due to metal load resulting in disruption of fish and algae growth (Sengupta, 1993; Sams and Beer, 2000; Salonen et al, 2003; Bell and Donnelly, 2006).

In the case of Britain, most coal field areas have been closed and mine water pumping has stopped. As a result, the emission of ferruginous effluents and *acid mine drainage* from the various exits to mines due to groundwater rebound are two of the most remarkable effects of coal mine closure (Bell and Donnelly, 2006). In Adak, located in the Vasterbotten district in northern Sweden, surface water, sediment and soil samples contain higher concentrations of As, Cu, Fe and Zn, compared to the target and intervention limits set by international regulatory agencies (Bhattacharya et al. 2006).

Studies of both Song et al. (1997) in Daesong coal mine, Keumsan in South Korea and Sams and Beer (2000) along the Allegheny and Monongahela rivers in the US including their sub-basins revealed the extent of polluted area by *acid mine drainage* due to upward trends in sulphate concentrations. These trends appear to be related to increase in coal production.

Klukanova and Rapant (1999) showed that waters draining freely from the mines transport large amounts of toxic elements into surface streams which contaminate broader surroundings in the Handlova–Cigel brown coal district, Slovakia. The results from monitored localities indicated that long-term mining activities adversely influence the environment. Although during the past decades or centuries many of these effects may have been reduced or eliminated, such as in old dumps that were covered by vegetation and where their toxic elements washed out to become part of present-day environment. However, many past mining activities cause environmental problems even today, and they must be mapped and monitored.

In order to have detailed knowledge on the extent of impacts of opencast mining, site assessment is necessary and various kinds of investigations should be explored in order to choose the best technique for the environmental reclamation. Various analysis techniques, sampling and modeling schemes have been proposed and applied by researchers and according to Cuccu (2002), judgmental sampling, systematic and regular grid sampling, simple random sampling, stratified sampling, ranked set sampling, composite sampling can all be used as sampling method, and the techniques to achieve sampling operations may differ in function of number of sampling area and geometric features of sampling location.

Navarro et al. (2004) have carried out field and laboratory studies in order to investigate soil contamination derived from past mining activity in the Sierra Almagrera district in southeast Spain. According to the study, the tailings, soil and sediment samples that were collected showed high concentrations of Ag, As, Ba, Cu, Pb, Sb and Zn when analyzed.

Navarro et al. (2008) evaluated the dispersion and influence of soluble and particulate metals present in the materials from an abandoned mine, Cabezo Rajao, in Spain. Tailings and soils were sampled and analyzed for pH, EC, CaCO3, grain size, mineralogical composition and heavy metal content, while water samples were collected and analyzed for pH, EC, soluble metals and salts. A total of eighteen sampling stations were selected from Cabezo Rajao mining site, to be representative of the different soils or waste material types present at the site. Solid samples were air dried and sieved to < 2 mm for general analytical determinations. Equivalent calcium carbonate was determined by the volumetric method using a Bernard calcimeter. Textural analysis was performed after dispersion of the fine soil and by combining extraction by Robinson pipette and sieving, and the mineralogical composition of the samples was determined by X-ray diffraction analysis.

In the study conducted by Jun-bao et al. (2002) in the Fushun coal mine, the northeast China, the spatiotemporal variation of heavy metal element content in reclamation soil was studied and grid method was used in order to sample covering soil at the test field. The soil samples were taken at different locations, including three kinds of covering soil, three different depths of soil layers and four different covering ages of covering soil.

Komnitsas and Modis (2006) aim to map As and Zn contamination and assess the risk for agricultural soils in a wider disposal site containing wastes derived from coal beneficiation in coal mining region of Tula, south of Moscow, Russia. Geochemical data related to environmental studies show that the waste characteristics favor solubilization and mobilization of inorganic contaminants and in some cases the generation of acidic leachates. 135 soil samples were collected from a depth of 20 cm using a 500 m x 500 m grid and analyzed by using geostatistics under the maximum entropy principle in order to produce risk assessment maps and estimate the probability of soil contamination. The samples were oven dried, sieved, ground, dissolved in aqua regia and analyzed for 23 inorganic elements by atomic absorption spectrophotometry.

All types of opencast mining have serious impact on all landscape components and functions, leading to significant alteration of the original landscape, which is actually a subcategory of cultural landscape. Once mining operations start, the landscape development in progress is disturbed, the original ecosystems are removed, the topography is significantly altered, the basic ecological relations are unchangeably disrupted, and biodiversity is decreased. These factors consequently lead to total ecological destabilization, elimination of the aesthetic values and decrease in the recreational potential of the landscape. Therefore, post-mining landscapes are often called *"landscapes without a memory"*, which gives landscape architects one of the few opportunities in order to create a new landscape that will rapidly improve the visual quality of a region. (Sklenicka et al. 2004; Sklenicka and Kasparova 2008).

As a result of the aforementioned changes in the ecosystem, disturbance on nourishment and energy flows is inevitable, which mostly leads to devastation of the ecosystem. When viewed in the landscape planning perspective, landscape evaluation can be considered as a management tool and the first thing to consider is to decide for which purpose the landscape will be used. Then, the implementation of reclamation should be carried out by taking the basic rules of ecology into account. In case of having connected habitats, a small portion of land may serve as a healthy ecosystem. The impacts of mining on the social and environmental structure are mostly long-term and closely associated with the social level of the local society (Gillarova and Pecharova 2009).

RECLAMATION OF OPENCAST MINE SITES

It's crucial to make a mine disturbed land environmentally stable in order to transfer an unpolluted environment and natural resources to the next generations. However, when a demolished land is left with its own, it may take years and years to recover and reach an ecological balance. During this period, these types of lands need human hand for reclamation and recovery. Therefore, post-mining reclamation works are those aiming to regain landscape's fertility, its ecologic, economic and esthetic values (Akpınar, 2005).

Basic Terminology

There are different terms that have been used for reclamation such as *rehabilitation, restoration* and*recultivation*. Whereas these terms are mostly used interchangeably, there are obviously some fine differences in meaning.

Restoration is used as "the act of restoring to a former state or position or to an unimpaired or perfect condition". To restore means "*to bring back to the original state or to a healthy or vigorous state*". This usage implies returning to an original state and to a state that is perfect and healthy. On both sides of the Atlantic, the word is used in that way. Rehabilitation is "*the action of restoring a thing to a previous condition or status"*. This may sound similar to restoration; however, there is little or no implication of perfection. In common usage, something that is rehabilitated is not expected to be in as original or healthy state as if it had been restored. Remediation is the act of remedying. To remedy is *"to rectify, to make good"*. There is more emphasis on the process rather than on the endpoint reached. Reclamation is used particularly in Britain but also in North America. It is defined as *"the making of land fit for cultivation"*. However, to reclaim is defined as *"to bring back to a proper state"*. This definition does not imply returning to an original state but rather to a useful one. Replacement is, therefore, a possible alternative option. To replace is *"to provide or procure a substitute or equivalent in place of"* (although an alternative meaning is to restore). Mitigation is a word often used when restoration is considered. It is important to note that it is nothing to do with restoration. To mitigate means *"to appease or to moderate the heinousness of something"* (Bradshaw, 1996).

Three categories of remedial treatment have been defined by the National Academy of Sciences, America: *"Rehabilitation:* The land is returned to a form and productivity in conformity with a prior land-use plan with a stable ecological state that does not contribute substantially to environmental deterioration and is consistent with surrounding aesthetic values."; *"Reclamation:* The site is hospitable to organisms that were originally present or others that approximate the original inhabitants."; *"Restoration:* The condition of the site at the time of disturbance is replicated after the action.". According to these definitions: i) *rehabilitation* usually allows the greatest flexibility in future land use and incurs the least cost; ii) *reclamation* means that the pre- and post-disturbance land uses are nearly the same; and iii) *restoration* allows no land use flexibility and results in the greatest cost (Sahu and Dash, 2011).

In British terminology, *"restoration* means the return of newly mined land to post-mining productivity, whereas *reclamation* means the recovery of derelict land (abandoned industrial land including that from mining) to usefulness. American usage of the word *restoration* has caused it to mean a strict replication of conditions existing before mining." (Saperstein 1990).

According to Del Tredici (2008), *"restoration* has inherent assumptions stating that it is both possible and desirable to establish some portion of the original ecological conditions of a site. People in favor of following strict restoration guidelines have to answer two very difficult questions: i) to what former time period should the site be restored? And ii) how should one deal with the imponderable environmental changes affecting the site? On the other hand, *reclamation*, also referred to as*revitalization*, assumes that there is no ecologic time travel to an earlier state of the site. Instead, to minimize the negative impacts of the site on the surrounding environment and to maximize its aesthetic and ecological functionally are the main objectives of reclamation projects, which are usually large scale and heavily disturbed".

Aim, Importance and Necessity of Reclamation

A rational reclamation objective should not only aim to create a permanently stable landscape that is both aesthetically and environmentally compatible with surrounding undisturbed lands, but also take into consideration aesthetics, intended use, and versatility when shaping the land in order to construct a land resource with both maximum feasible utility and versatility for future generations. Even though the approximate original contour as a minimum condition is generally required by reclamation regulations, there can be cases where variance from that is allowed as long as desirable results are guaranteed (Jansen and Melsted 1988; Sengupta 1993).

Within the frame of remediation of a contaminated land, either the minimization of actual or potential environment threat, or the reduction of potential risks to acceptable levels are the main goals, which can be accomplished by applying one or more of the following (Wood 1997; 2001):

- elimination of the hazard by removal or treatment/modification of the contaminant,
- control of the hazard by isolation or separation of the contaminant,
- interruption of the pathway of contaminant movement and exposure,
- protection or removal of the receptor (essentially involving an interruption of the pathway).

When viewed in mine reclamation perspective, the fundamental objectives are given as (Cao 2007):

- to eliminate health and safety hazards (i.e., removal of all facilities and structures threatening human health and safety),
- to restore impacted land and water resources (i.e., progressive re-vegetation and stabilization of residues to reduce potential of acid mine drainage or water contamination),
- to eliminate off-site environmental impacts (i.e., cleaning up sites to conform to the community's surrounding landscape),
- to ensure that post-mining land has a feasible self-sustaining future with respect to both environmental and socio-economic benefits (i.e., developing publicly owned land for recreation, historic purposes, conservation purposes, or open space benefits, or for constructing public facilities in communities),
- to encourage better use of energy and natural resources, and to guarantee sustained mining operations.

Mining and land development are closely linked in the dynamic and integrative process addressed by a range of environmental, production, aesthetic, land use, and economic issues related within the reclamation planning objectives. This process, whose outline is briefly given below, starts at the opening of a mine operation and terminates at the closure of the mine, which may take five to fifty years (Bauer, 2000):

- build a mine environment compatible with neighboring land uses during the whole mine operation,
- maximize access to aggregate resources on the site,
- use all unique deposit features created by the mining operation in shaping new landscapes,
- employ non-aggregate earth materials such as overburden, clay deposits, and mine waste in building and shaping land forms,
- use available earth moving equipment and earth moving procedures efficiently for reclaiming the mine site, without interfering with ongoing mining operations,
- develop a coordinated and sequential program of mining, earth moving, land shaping, and landscaping to ensure that lands are prepared for development as mining progresses through the deposit.

Methods and Techniques for Reclamation

The process of choosing the most appropriate technique for the reclamation is often a painstaking task, and many economic and operational parameters (i.e., process applicability, effectiveness and costs, process development status and availability and operational requirements) should be taken into account, as well as several additional factors such as process limitations, monitoring needs, potential environmental impact, health and safety needs and post-treatment management requirements (Wood 1997). Additionally, large extend of areas within the mining and industrial structures (i.e., traffic network, electricity grid, pipelines, canalization streams, storage areas, industrial parks etc.) also increase the complexity of rehabilitation works, and restrain the possibilities (Prikryl et al. 2002).

In reclamation of post-mining landscapes, there are generally two basic motives: determinism and contingency, and the processes related with the former are mostly considered. Several factors associated with the latter have also significant role on the success of the reclamation, and these factors are often unpredictable and can be grouped in four categories: i) Initial conditions (natural climate and topography, type and abundance of topsoil); ii) Natural perturbations (droughts, extreme rainfall events, frost periods, pests); iii) Influence of the surrounding ecosystems and people (runoff and sediment flows, grazing, hunting, land uses); and iv) Human contingencies (modification/intermittence of mining operations; mistakes in the performance of reclamation works; changes in legal rules, etc.). Additionally, it is necessary to consider the reclaimed areas as open ecosystems that interact with their surrounding environment, so landscaping schemes and reclamation work must be included in any proposal for the development of a mine that broadly evaluates the impacts of open mine sites on local residents, the landscape and the environment. The following points should be considered to improve the performance of opencast mine reclamation (Bell and Donnelly 2006; Ibarra and de las Heras 2005):

- simultaneous integration of mining and reclamation activities to optimize the opportunities offered by mining operations,
- interactive development of reclamation projects by all actors and to have a consensus on the final objectives for the reclaimed areas,

- specific research to acquire detailed knowledge about the reference ecosystem in order to adopt the general protocols for reclamation to local conditions,
- plan for monitoring and survey to check, improve, or redirect the applied practices.

For achievable and sustainable reclamation, detailed information on various important factors, classified as either natural or cultural, is required during the post-mining land use planning, and they are listed in Table 1 (Ramani et al. 1990):

Naturally, not every mine has the same motives and methods for site rehabilitation, and it is important to point out that it is not feasible to restore all mine sites due to economic and operational considerations. However, even though disturbed by mining activities, all post-mining lands eventually inherit some economic, recreational and esthetic potential. Hence, discovering the unique potential of mined land and choosing appropriate methods and measures, which actually form the core of reclamation, are necessary for the successful transformation of this potential into a sustained capability. In order to obtain satisfactory results in reclamation, special attention must be paid to the post-mining use of the land and its potential functions (i.e., pasture, hayland, recreational areas, wildlife habitat, wetlands, fishing ponds etc.), together with the implementation of environmental conservation and land reclamation programs to minimize the negative environmental effects (Cao 2007; Kavourides et al. 2002; Saperstein 1990).

Table 1: Required information for reclamation and postmining land use planning (Ramani et al. 1990)

NATURAL FACTORS	Terrestrial ecology
Topography	Natural vegetation, characterization, identification
Relief	of survival needs
Slope	Crops
Climate	Game animals
Precipitation	Resident and migratory birds

wind-airflow patterns, intensity	Rare and endangered species
Humidity	Aquatic ecology
Temperature	Aquatic animals-fish; water birds, resident and
Climate type	migratory
Growing season	Aquatic plants
Microclimatic characteristics	Characterization, use, and survival needs of aquatic
Altitude	life system
Exposure (aspect)	
Hydrology	CULTURAL FACTORS
Surface hydrology	Location
watershed consideration	Accessibility
flood plain delineations	Travel distance
surface drainage patterns	Travel time
amount and quality of runoffs	Transportation networks
Ground water hydrology	Size and shape of the site
ground water table	Surrounding land use
aquifers	Current
amount and quality of ground	Historical
water flows	Land use plans
recharge potential	Zoning ordinances
Geology	Land ownership
Stratigraphy	Public
Structure	Industry
Geomorphology	Private

Chemical nature of overburden	Type, intensity, and value of use
Coal characterization	Agriculture
Soil	Forestry
Agricultural characteristics	Recreational
Texture	Residential
Structure	Commercial
organic matter content	Industrial
moisture content	Institutional
permeability	Transportation/Utilities
pH	Water
depth to bedrock	Population characteristics
Color	Population
Engineering characteristics	Population shift
shrink-swell potential	Density
Wetness	Age distribution
depth to bedrock	Number of households
erodibility	Household size
slope	Average income
bearing capacity	Employment Educational
organic layers	levels

Since it is inevitable to have various mutual associates (i.e., companies, state and local agencies, as well as special interest groups and general public) in the planning of surface mines, the major objective is to preserve or enhance the long term use of the land within an integrated mining, reclamation, and land use planning concept that accounts for the interactions that must take place between the various levels of land use planners. A sample framework of such a plan is illustrated inFigure 1 (Ramani et al. 1990).

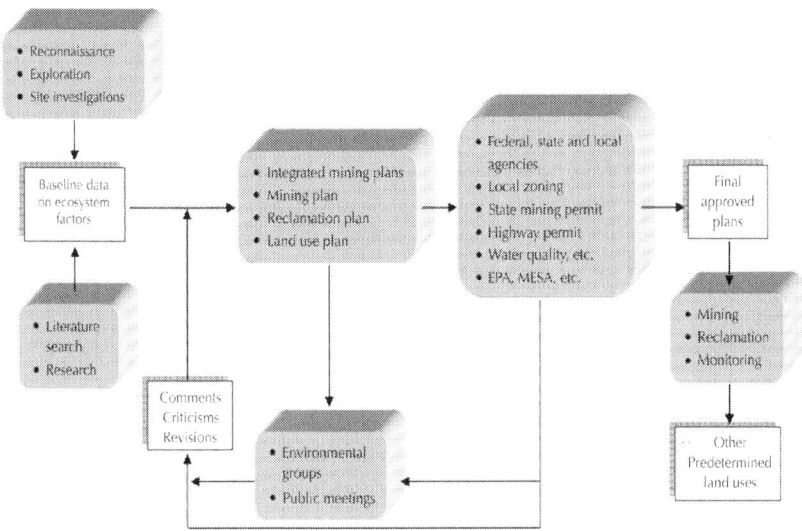

Figure 1: Process of the integration of surface mining, reclamation, and land use planning (Ramani et al. 1990).

Rearrangement and rehabilitation works, which may be either to remove the visual effects of an existing mine site or to reduce the impact of a new mine site to a lowest degree, should be planned before starting operation and carried out in parallel with mining activities. By this way, reclamation can be implemented in a more economical way with minimum cost. During the planning of rehabilitation works, research for land use and purpose of use for the reclaimed land are crucial (Akpınar et al. 1993), and this type of landscape planning should fulfill some or all of the following general conditions (Fanuscu 1999; Görcelio lu 2002):

- During mining operations:
- to minimize the visual impact at lowest possible degree,
- to take the necessary landscape planning measures against noise and dust contamination.
- Subsequent to mining operations:
- to carry out an effective and economic rehabilitation in order to have an efficient post-mining utilization,
- to reform the land in accordance with the final scope of use within the frame of available resources,

- to rehabilitate the lower layer material, which has been dug up and is inappropriate for vegetation,

to take replantation and post-mining land use issues into account.

According to Akpınar et al. (1993), rearrangement and rehabilitation works on degraded areas due to mining activities are carried out in four main steps: i) *post-mining land use planning*; ii)*rearrangement* within the frame of existing land use plan (excavation, dumping, water regime control, removing and laying out of top soil separately etc.); iii) *rehabilitation* (biological reclamation); and iv)*monitoring* and *maintenance*.

One of the main aims of reclamation is to restore the land use capability of disturbed landscape; within this context, reclamation planning is necessary and strictly related to *land use planning*. Substantial deformation of the topographic structure of the landscape, loss of fertile top soil, detriment in the flora and fauna; reduction of such negative effects to a minimum level or complete removal of them are achieved by *landscape use planning*, which is the first stage of the landscape restoration studies.

Landscape use planning, briefly, is to investigate a landscape in point of different aspects and research its availability for the proposed purpose of use. Such plans ensure an optimum utilization of resources by either preservation of environmental values or reducing harmful effects. Re-establishing the balance between ecology and economy in order to decrease the inevitable environmental problems at a minimum level caused by mining, reconstructing the disturbed ecosystem and introducing the possible new uses according to the needs of dwellers are among the main concerns of *landscape use planning*. These studies are part of the reclamation work and begin with planning of all mining activities, then continue during the whole production process. At the beginning, preliminary decisions about the post-mining land use of region are made. This initial plan constitutes the base for detailed decisions to be made later and provides a chance for preliminary evaluation (Akpınar et al. 1993).

Potential future use of the post-mining lands basically depends on the nature of the land, soil conditions, and communal structure of nearby surrounding to be rehabilitated by technical, biological, agricultural means or forestry applications. The followings are the potential land use types that follow successful land reclamation (Görcelio lu 2002; Topay et al. 2007; Tshivhandekano 2004):

- The original land use,
- Afforestation, forestry,
- Agriculture,
- Nature conservation and wildlife,
- Hydrology,
- Recreation,
- Site improving,
- Special reserve,
- Settlement or industry,
- Solid waste or rubble storage area.

Rearrangement includes excavation and dumping according to the planning, stable design of dump sites and chamfers with proper slope and elevation, laying out of top humus layer and fertile soil right beneath it either directly or later, grading, drainage and water regime control, constructing surrounding drainage channels against floods, and constructing infrastructure and road network; whereas*rehabilitation* comprises improvement of soil conditions and re-vegetation on topographically graded lands (Akpınar et al., 1993).

Factors related to soil and climatic conditions play significant role in the reclamation of post-mining lands. Although it is easy to modify soil factors, it is practically impossible to manipulate the climatic factors except for those related to moisture (i.e., irrigation for water-deficient periods and drainage for water excess periods). Since the climatic adaptation of plants to a certain specific region is one of the major concerns for the fulfillment of reclamation objectives, a special attention must be paid to climate when selecting plants for erosion control and targeted land uses (Powell 1988).

Since there is an urgent need for soil reconstruction and restoration of productive and functional soil plant systems on abandoned and degraded opencast mine sites, soil improvement is an indispensable stage of any reclamation process, where geological substrate, slope and type of reclamation are the key determinants. The whole process consists of a sequence of interrelated stages: i) application of additives, ii) spreading and defraying organic materials, and iii) fertilizing crop rotation. It has also be noted that Mg and Ca amounts, absorption capacity, and available humus forms of the soil horizons should be carefully analyzed in deciding which type of plant is more appropriate (i.e., deciduous or coniferous) (Hendrychova 2008).

Vegetation cover has some significant functions on post-mining landscapes in different ways by i) modifying the surface characteristics, ii) controlling the erosion and reducing slope failures, iii) lessening stream sedimentation, and iv) restoring the beauty and productivity of the land (Schor and Gray 2007; Hutnik and McKee 1990). So, in order to reduce the probability of negative consequences, selection of suitable plant material, which may be either native or introduced plant species, is critical. Additionally, the planners should not only take site specific conditions under consideration, but also pay special attention to those points during vegetation establishment, as will be explained now compactly (Hutnik and McKee 1990; Powell 1988):

- it should be considered as an integral part of the mining and reclamation process,
- appropriate methods should be chosen according to the aim of re-vegetation and the plant species used,
- local variations in climate, geology, and soils should be considered.

Basic knowledge about both biotic and abiotic factors, as well as ecological processes is necessary to reduce the time period needed for creating the favorable soil characteristics required for prosperous biological reclamation. The properties of the reconstructed soils should be analyzed since the structure of future ecosystem highly depends on physical and chemical soil characteristics, which directly affect the amounts of available resources (i.e., nutrient levels), initial species establishment, and long-term successional trends (Hendrychova 2008).

After a proper *rearrangement* and *rehabilitation* work, an additional time is needed to ensure a fertile use of land. At this stage, *monitoring, maintenance* and *controlling* of many environmental and ecologic parameters (i.e., water quality, drainage, vegetation growth, soil condition, erosion etc.) closely associated with the restoration site are essential to improve the quality of the restoration (Akpınar 2005).

Mining activities definitely have long-term impacts on terrestrial ecosystems: i) land degradation, ii) deforestation, iii) loss of fertile topsoil, iv) change in topography and hydrologic conditions, and v) pollution of usable surface and ground water (Tören 2002). So, monitoring and management of post-mining environment are necessary to evaluate the environmental impacts and long term behaviors of post-mining landscapes, and they should be handled with in the perspective of a well-planned environmental policy. Besides, even during any remediation process, adequate quality control measures are also needed to ensure that the methodology conforms to specification or that treatment targets have been achieved. In many cases, environmental monitoring is required while remediation is still in progress. These objectives naturally imply the utilization of scientific methodology, particularly, when field data is unavailable or insufficient (Hancock et al. 2006; Wood 1997).

The basic principles of the environmental management policy for reclamation are given as (Kavourides et al. 2002):

- knowledge of the local environmental conditions,
- selection of the proper methods and techniques of land reclamation,
- general land-planning for the areas under reclamation (land use map),
- systematic realization of the environmental protection and restoration programs according to the environmental terms determined by the Ministry,
- monitoring and evaluation of the environmental restoration results by geographic information systems (GIS).

Reclamation studies often require the integration of multi-source data acquired by different sources of diverse technical and operational characteristics (Kyzeridi et al. 2002). Such data is mostly in both time and spatial domains. So, use of GIS incorporated with remote sensing (RS) technologies provides a suitable platform for the monitoring and the management of reclamation, since it offers unique capabilities for editing, managing, analyzing and automating different kinds of spatial data required for decision making (Bruns and Sweet 2004; Chevrel et al. 2001; Chevrel et al. 2002; Smyth and Dearden 1998; Ganas et al. 2004).

GIS-based decision support systems have many potential applications in reclamation: i) to derive computer-based landscape evolution models for better understanding of geomorphic landscape process in reclamation (Hancock 2004), ii) to evaluate the future development of terrestrial ecosystem under the extreme environmental conditions of post-mining landscapes (Hüttl 1998), iii) to reduce the cost of spoil handling during mining and reclamation (Harwood and Thames 1988),

iv) to detect reclamation sites and to measure the impacts of increasing land degradation (Gorokhovich et al. 2003;Hladnik 2005), and v) to increase sophistication of mining industry in rehabilitation practices (Hancock 2004).

Evaluating the Success of Reclamation

Restoration of a landscape disturbed by opencast mining operations is mostly viewed in technical or economic perspectives only. Even though the public focused only on the forestry and agricultural aspects of restoration previously, there has been a recent interest in nature conservation and recreation. In order to restore ecological, hydrological, aesthetic, production, recreational and other functions of the post-mining area, a sustainable land use development plan should be prepared through a holistic approach (Sklenicka and Kasparova 2008).

Three basic goals that any restoration plan should reach are given as (Powell 1988):

- stabilization of newly reclaimed lands against accelerated wind and/or water erosion,
- development of target specific re-vegetation programs,
- achievable and sustainable land use by enforcing certain minimum performance standards.

The evaluation of restoration success is a tough issue, since it strictly depends on the character of the post-mining land, inherent features of ecological species involved, and the main objectives of the restoration operation. According to Pecharova et al. (2011), the process should allow spontaneous succession, or use technical restoration by sowing or planting target species and restoring or improving the site conditions.

The *Society for Ecological Restoration International* addresses the same issue by taking 9 ecosystem-related parameters under consideration to measure the restoration success (Hendrychova 2008):

- similar diversity and community structure in comparison with reference sites,
- presence of local species,
- functional groups necessary for long-term stability,

- capacity of the physical environment to sustain viable populations,
- regular functioning,
- integration with the landscape,
- removal of potential threats,
- resilience to natural disturbances,
- self-sustainability.

Interdisciplinary Dimension of Reclamation

All mining operations, due to their nature, have negative impacts on the cultural landscape. And opencast mining activities are not an exception, as they drastically change the former dynamic equilibrium of the landscape, leading to the formation of new ecosystems. Only the vegetation establishment by itself is not a proper approach. Instead, sustainable establishment of new ecosystems in the post-mining areas should be seen as an interdisciplinary challenge, in which the active participation of both science and society is highly required (Hüttl and Gerwin 2005).

The negative visual impact of the mining sites unavoidably lowers the aesthetic value of the landscape and its surroundings. So, post-mining landscape planning and rehabilitation activities should strictly consider the previous aesthetic characteristics of the land and their future development within an interdisciplinary approach (Sklenicka and Kasparova 2008).

As important as the interdisciplinary dimension of the issue, all mutual associates such as stakeholders, state officials and laws, environmental groups, engineers, landscape architectures, ecological experts, soil and social scientists should be "creatively" involved in the planning phase (Miao and Mars 2000;Saperstein 1990). By this way, significant social and environmental gains can be obtained by improving the conditions of post-mining landscapes through a relatively small investment (Garavan et.al. 2008).

When building up such a "project team", scale of the project and its complexity, special issues, and the clients' demands are the key factors to get the best results. For projects that require permitting and the preparation of mining and reclamation documents, landscape architects, mine operators, geologists, hydro-geologists, and civil engineers are possible candidates for the team (Bauer 2000).

The involvement of landscape architects in the industry has increased steadily and their role has gone far beyond the "traditional boundaries" of the profession; commonly and mistakenly thought as basic beautification and site planning duties. Based on their education and experiences, landscape architects now can easily deal with more complex sequential mining and reclamation plans, including site analysis, site and land use planning, visual analysis, grading, zoning, re-vegetation, slope stabilization, etc. They also play active roles in the permitting, regulatory, environmental assessment and community relation processes. The form, the function and the purpose of post mine landscape planning should be considered within earth science issues by landscape architects. By this way, they can approach the issue with more systematic and comprehensive manner. Landscape architects should aim to develop integrated multi-scale design approaches not only to be an equal partner in the planning process, but also to be in a position to direct the project team and undertake the responsibility for the success of reclamation planning (Arbogast 2008; Bauer 2000).

For a landscape architecture, a working knowledge on the following three points is essential for the understanding and success of reclamation planning services (Bauer 2000): i) components of mining processes associated with reclamation, ii) geologic complexities and structures within each aggregate deposit, and iii) mechanics and procedures for incorporating the mining procedures with reclamation activities. Table 2 overviews the mine reclamation studies within a multidisciplinary point of view.

Table 2: The areas of technical expertise essential in pre-mining (Adapted from Ramani et.al. 1990; im ir et al. 2007)

Mine Planning Phase	Planning Activities	Areas of Specialization
Legal requirements analysis	Identification of regulatory constraints related to land use	Land use planner Attorney or paralegal specialist

Land and reserve acquisition	Prepare land use / land cover maps Prepare land ownership map	Land use planner Landscape architect Photogrammetrist/ cartographer Plant biologist Photogrammetrist/ cartographer Surveyor
Market development	Check market potential of site	Geographer Transportation engineer Land use planner
Financial evaluation	Check if land development potential of the site will justify reclamation to a higher, more costly land use	Engineering economist Land use planner Real estate specialist Fiscal planner
Coal beneficiation studies and plant design	Determine the impact of waste disposal on the Postmining uses of land	Mineral processing engineer Environmental engineer Landscape architect Agronomist Geologist Hydrologist
Environmental impact studies	Evaluate the impact mining will have on the site with respect to capability and productivity	Mining engineer Environmental engineer Forest engineer Agronomist Geologist Hydrogeologist Terrestrial ecologist Plant biologist Agricultural engineer Archeologist Landscape architect Land use planner Social scientist
Preliminary mine planning	Preliminary identification of postmining land uses	Mining engineer Land use planner Landscape architect Agronomist Engineering economist
Permits acquisition	Land use information and postmining land use plan	Mining engineer Land use planner Environmental engineer Agronomist
Administrative detail analysis	Submitted / approval of final land use plan	Agricultural engineer Hydrogeologist Plant biologist Engineering economist

Detailed mine planning	Detailed land use plan design	Land use planner (specifically landscape architect) Mining engineer Environmental engineer Civil engineer Agricultural engineer Agronomist Hydrogeologist Plant biologist Engineering economist

Case Studies of Reclamation

In the 20th century, rapid developments and new innovations in the technology and the machinery used in the mining industry have changed the whole face of landscape modification in all large mining districts of the world. Common problems in such post-mining areas are the increase in water surface area and the acid mine drainage, leading to severe site conditions. Hence, prior to any reclamation study, site specific conditions due to previous mining activities should be taken into account, and the plans toward sustainable ecosystem development should be prepared accordingly. Additionally, soil fauna, mechanisms of plant succession and site specific biocoenosis should be investigated thoroughly for the successful establishment of terrestrial ecosystems on post-mining sites (Hüttl and Gerwin 2005).

Flambeau Mine, located in Wisconsin, is one of the prominent examples for the application of sustainable development principles and implementation of twenty first century materials and engineering technology to reclamation of post-mining landscapes. The implementation of sustainable development at Flambeau Mine has four main pillars: i) economic prosperity, ii) environmental protection, iii) social and community well-being, and iv) governance. The design is based upon a collaborative approach from overlapping disciplines. In addition to traditional engineers and architects, community planners, transportation planners, biodiversity specialists, energy efficiency specialists (e.g. green building) and landscape architects all contribute to the master design. The key to a successful design is to meld the wants and needs of the community with the various ideas and designs from the design team (Cherry 2008).

The study was designed, constructed, operated, and reclaimed in the 1990s. Reclamation of the site began during the fall of 1996 with the initiation of sequential backfilling of the open pit, which was substantially complete by the fall of 1997. During 1998, the contours of the site were reestablished, topsoil replaced, wetlands constructed, and seeding and planting were initiated. The majority of seeding and planting was completed by year and 1999. Additionally the design constructed hiking, biking and equestrian trails for public recreational use. The pre-mining, active mining, and reclaimed site are shown in Figure 2 for a chronological comparison (Fox 2002 ; Cherry 2008).

Figure 2: Flambeau Mine Site: a) before mining (1991), b) during mining (1996), and c) after mining (2002) (Fox 2002).

The reclamation of the Flambeau Mine has included (Flambeau Reclaimed 2012):

- returning the site to its original approximate contours,
- planting clusters of trees to attract and support wildlife habitats,
- creating and restoring over 10 acres of wetland on site,
- creating over 120 acres of grassland habitat,
- constructing four miles of trails for non-motorized recreational activities.

The completion of surface contouring and return of the topsoil in 1998 were followed by the planting of native plant species necessary for the creation of prairie grasslands, woodlands and wetlands. In order to monitor and evaluate the success of the reclamation, 300 locations were randomly selected across the reclaimed Flambeau Mine. At these locations, necessary studies are performed each year in order to observe whether the performance standards (i.e., vegetative cover, planted native species, diversity and woody species survival) are met. In 2001, all necessary standards were met at the reclaimed mine site, which allowed the submittal of the *Notice of Completion to the Wisconsin Department of Natural Resources*. Recent surveys show a fully utilized wildlife at the reclaimed site, which provides unique and critical habitat, particularly for grassland bird species (Flambeau Reclaimed 2012).

Another outstanding example is Jarrahdale in Australia. The mining operations for bauxite at Jarrahdele started in 1963 and continued until 1998. During this period, over 160 million tons of ore was mined. Site rehabilitation studies continued for another 3 years. By 2001, all mined areas, haul roads and building sites were completely rehabilitated (Figure 3) (Alcoa 2012).

Figure 3: a) The original Jarrahdale crusher circle before its closure in 1998, and b) the same crusher circle site at Jarrahdale, after rehabilitation has been completed (Alcao 2012).

As shown in Figure 4, reclaimed mine sites are returned to productive use in a variety of ways that will serve for future generations.

Figure 4: A site reclaimed by Starvaggi Industries in West Virginia is developed into the Star Lake Amphitheater: a) post-mining landscape, b) after the reclamation (Mineral Information Institute 2012).

In Turkey, reclamation of abandoned mine sites is generally carried out in the form afforestation (Figure 5 and 6).

Figure 5: Afforestation operations in ile, stanbul (ile Forestry Operation Directorate 2012).

Figure 6: Afforestation in A açlı, stanbul (Kutorman 2012).

Rehabilitation and restoration operations in most of the abandoned coal mine areas are conducted by Turkish Coal Enterprises (TK). According to General Directorate of Turkish Coal Enterprises (2011), between 1991 and 2011, nearly 7.3 million trees in various species [stone pine (*Pinus pinea*), black pine (*Pinus nigra*), red pine (*Pinus brutia*), cypress (*Cupressus* sp.), cedar (*Cedrus* sp.), horse chestnut (*Aesculus hippocastanum*), black locust (*Robinia pseudoacasia*), tree of heaven (*Ailanthus altissima*), oak (*Quercus* sp.), maple (*Acer* sp.), ash (*Fraxinus* sp.), etc.] were planted on 4455 hectares of post-mining lands in various establishment directorates of TK (Figure 7).

Figure 7: Afforestation of soil waste dumps by the establishment directorates of TK (General Directorate of Turkish Coal Enterprises 2011).

Legislative and Regulatory Issues in Mine Reclamation

Since the late 20th century, reclamation has been widely accepted by both developed and developing countries as a desirable and necessary remedy in order to: i) reestablish the environmental conditions in post-mining landscapes at an acceptable level, and ii) increase their economic value to an optimum level (Cao 2007).

The law plays a critical role in reclamation of the post-mining landscape. It does not only define the legal status of the issue, but also reveals the outlook and approach of individual governments, which differ significantly in their attempts to mitigate the effects of mining disturbance.

For many of the developing countries, mining has a significant contribution to economy, which often puts a certain pressure on policy makers in order to establish an appropriate balance between national economic growth and environmental protection. Generally speaking, developing countries do not have strict environmental regulations and effective enforcement programs, and they usually address the issue within mining and environmental acts, or related national laws. Additionally, these countries mostly consider the reclamation and pollution control after the mine operations end (Cao 2007).

On the other hand, the approach in developed countries is more comprehensive and they have more stringent and effective regulations. Besides, restoration is regarded as a continuous process during mining, and mining companies have to prepare detailed environmental management plans and use expensive environmental technologies.

At this point, it would be wise to basically give examples of legislations related to mining and reclamation in several developed countries. The situation in Turkey is also overviewed briefly.

USA

With both the foundation of the *Soil Conservation Service* in the early 1930s and increasing local and state concerns about the degradation of land due to surface mining operations, protection of land resources became a publically important issue after World War I. This movement evolved into surface mine legislation, first in West Virginia in 1938, and then, in Indiana (1941), Illinois (1943), Pennsylvania (1945), and in Ohio (1947). Parallel to the increase in surface mining activities due to the energy crisis in the 1970s, the protection of environment gained more public interest (Doll 1988).

In the first half of the 1970s, many states asked mine operators to provide methods for certain mine operations and requirements for reclamation. During the late 1970s and early 1980s, more compulsive regulations were imposed by the states with coal mining activities and the federal government. In this period, under the influence of Congress and pending legislation, public education campaigns by local mining associations and new research efforts by the industry to reduce the economic impact of legislated reclamation gained speed. Today, all surface mining in the US is regulated by federal or state laws (Harwood and Thames 1988).

The reclamation for the surface effects of coal mining activities (including underground operations) on public and private lands in the US is based on the *Surface Mining Control and Reclamation Act(SMCRA)* of 1977 (Micsak, 2008).

This principle law (i.e., *SMCRA/PL 95-87*) defines the federal standards for the reclamation of surface mine sites. Within the guidelines and regulatory procedures set by this law, the industry was pinned for the reclamation of surface-mined lands, which has led to major changes in mining practices and reclamation techniques. By this way, many surface-mined lands have been successfully reclaimed (Doll 1988).

Section 101(c) of *SMCRA* states: "Many surface mining operations result in disturbances of surface areas that burden and adversely affect commerce and the public welfare by destroying or diminishing the utility of land for commercial, industrial, residential, recreational, agricultural, and forestry purposes, by causing erosion and landslides, by contributing to floods, by polluting the water, by destroying fish and wildlife habitats, by impairing natural beauty, by damaging the property of citizens, by creating hazards dangerous to life and property by degrading the quality of life in local communities, and by counteracting governmental programs and efforts to conserve soil, water, and other natural resources." Section 101(e) of *SMCRA* says: "Surface mining and reclamation technology are now developed so that effective and reasonable regulation of surface coal mining operations by the States and by the Federal Government in accordance with the requirements of this Act is an appropriate and necessary means to minimize so far as practicable the adverse social, economic, and environmental effects of such mining operations." (Office of Surface Mining Reclamation and Enforcement 2012).

Canada

Each Canadian provincial government has the authority to make laws related to property, contracts, natural resources, employment, land use and planning, education, health care and municipalities. So, most laws in respect of commercial nature are enacted by provincial governments. Mining activities are mainly governed by the laws of the province or territory where a mine is physically located. Additionally, the federal government has overlapping jurisdiction in a number of areas such as taxation and the environment. The federal *Canadian Environmental Assessment Act (CEAA-2012)* constitutes the main legislative frame for all environmental assessment processes. It requires an environmental assessment when a federal authority proposes the mining project, provides financing or lands for the project, or issues certain permits or approvals for the project. In general, a federal environmental assessment is required for most major mining projects. Federal and/or provincial environmental impact assessments are required prior to commencing or expanding operations or even conducting exploration in order to decide whether or not a proposed mining project should proceed based on its environmental and social impacts. The government generally has the authority to require a public hearing and the discretion to accept a proposed mining project or reject it (Davies 2011).

Australia

The first Australian mining law dates back to 1851. Legal dimension of environmental issues associated with mine operations are defined within the various sections of *Mining Act* and the*Environmental Protection Act*, which was enacted in 1986. According to the act, any project proposal, which may potentially have a significant impact on the environment, is referred to the *Environmental Protection Authority*. The *Environmental Protection Authority* evaluates the proposal and prepares a report on whether the proposal should proceed. In relation to the minerals and the environment, four important points are always kept under consideration: i) assessment and recommendation on the environmental management related to exploration and mining proposal, ii) collaboration with the industry and the community on the environmental management of the mining industry, iii) compliance with environmental conditions during exploration and mining, and iv) cooperation with other governmental agencies in order to keep lands of high conservation under protection, and not to exclude land unnecessarily from exploration and development activity (Hunt 2009).

Germany

German mining law dates back to 1865, when the *Allgemeines Berggesetz (AGB)* was established. The first reclamation amendments to the mining law were enacted in 1929. Due to the increase in demand for coal after World War II, reclamation was ignored. However, beginning in 1950, reclamation efforts increased and new laws with more precise requirements were put into force in Germany (Knabe 1964). The act has been amended several times and was replaced in 1980 by the

Federal German Mining Act(*BGBl. IS. 1310*). This act was set into force in January 1982 and revised on December 9, 2006, through slight revision to provisions of *Article 11* (*BGBl. IS. 2833*) (Anderson 2012; Betlem et al. 2002).

United Kingdom

The main laws related to the mining and the environment in UK are i) *Coal Mines Regulation Act*(1908), ii) *Mining Industry Act* (1920), iii) *Coal Act* (1938), iv) *The Town and Country Planning Act*(Scotland) (1947), v) *Coal Industry Act* (1949), vi) *Mineral Workings Act* (1951), vii) *Mines and Quarries Act* (1954), viii) *Opencast Coal Act* (1958), ix) *Mines Act* (Northern Ireland) (1969), and x)*Environmental Protection Act* (1990) (Legislation.gov.uk 2012). English mining law operates primarily by public (administrative) law rather than by private (civil) mechanisms. The central administrative body is the *Coal Authority* and it was established under the *Coal Industry Act* (1994) during the privatization of the industry. There are lots of acts in the area of mining regulation; however, the *Coal Industry Act* (1994) and the *Coal Mining Subsidence Act* (1991) are the most pertinent ones (Betlem et al., 2002). *Environmental Protection Act* was amended by *Environment Act*in 1995, and *Part IIA* of this amendment defines a detailed framework for the identification and the compulsory remedial action for contaminated land (Legislation.gov.uk 2012).

France

The *French Mining Code* (*Code Minier*) was enacted on 21 April 1810. The old *Mining Code* was amended by *Law No. 94- 588* of 15 July 1994, which organizes existing case law and aims at a better protection of the environment, and can be seen as revisions to bring the *French Mining Code* in conformity with relevant European regulations. During the development of the French environmental law in the past three decades, mechanisms for financial sanctions for those causing environmental damage have been incorporated without proper coordination in enforcement. With the *Environmental Code* enacted in 1999 (*Code de l'Environnement*), a more coherent regime was aimed by the Government. The *Code* addresses to several environmental issues in more than 975 articles over six chapters, combining liability clauses (Betlem et al. 2002). The central government representatives (*préfets*) can legislate for promoting the conservation of the habitat of listed protected species, according to a decree adopted in 1977 for the implementation of the *Act* (Groombridge 1992).

Turkey

In Turkey, there have been several efforts to designate the principle legal guidelines for the reclamation of post-industrial landscapes. *"The Regulation on Reclamation of Lands Disturbed by Mining Activities"* is an important landmark for mine closure planning in Turkey. It basically aims to establish the basic requirements for this purpose, and was published on 14th of December, 2007, and amended on 23rd of January, 2010. According to this regulation, reclamation plans for mining projects must be appended to the *Environmental Impact Assessment* (*EIA*) reports. A summary of related laws and regulations is given in Table 3.

Table 3: The main laws and regulations related to mining and reclamation in Turkey (Official Gazette of Republic of Turkey 2012; Republic of Turkey Prime Ministry 2012; Republic of Turkey Ministry of Justice 2012; Chamber of Mining Engineers of Turkey 2012)

Laws and Regulations	Effective Date	Repealed by	Valid
General Hygiene Law No. 1593	1930	-	General Hygiene Law No. 1593 (1930)
Forest Law No. 6831	1956	-	Forest Law No. 6831 (1956)
The Constitution of the Republic of Turkey	1982	-	The Constitution of the Republic of Turkey (1982)
Environmental Law No. 2872	1983	-	Environmental Law No. 2872 (1983)
National Parks Law No. 2873	1983	-	National Parks Law No. 2873 (1983)

Regulation on Unhealthy Institutions	1983	Regulation on Unhealthy Institutions (1995) Regulation on Repealing of Unhealthy Institutions Regulation (2005)	Repealed (2005)
Mining Law No. 3213	1985	-	Mining Law No. 3213 (1985)
Regulation on Protection of Air Quality	1986	Regulation on Air Quality Assessment and Management (2008)	Regulation on Air Quality Assessment and Management (2008)
Regulation on Noise Control	1986	Regulation on Assessment and Management of Environmental Noise (2005; 2008; 2010)	Regulation on Assessment and Management of Environmental Noise (2010)
Regulation on Water Pollution Control	1988	Regulation on Water Pollution Control (2004)	Regulation on Water Pollution Control (2004)
Regulation on Control of Solid Wastes	1991	-	Regulation on Control of Solid Wastes (1991)
Regulation on Environmental Impact Assessment	1993	Regulation on Environmental Impact Assessment (1997; 2002; 2003; 2008)	Regulation on Environmental Impact Assessment (2008)

Regulation on Allocation of Forest Lands	1995	Regulation on Permissions in Forest Lands (2007) Regulation on Implementation of 17th and 18th Articles of the Forest Law (2011)	Regulation on Implementation of 17th and 18th Articles of the Forest Law (2011)
Regulation on Control of Hazardous Wastes	1995	Regulation on Control of Hazardous Wastes (2005)	Regulation on Control of Hazardous Wastes (2005)
Regulation on Control of Soil Pollution	2001	Regulation on Control of Soil Pollution (2005) Regulation on Control of Soil Pollution and Point-Source Contaminated Fields (2010)	Regulation on Control of Soil Pollution and Point-Source Contaminated Fields (2010)
Regulation on Noise	2003	-	Regulation on Noise (2003)
Regulation on the Implementation of Mining Law	2005	Regulation on the Implementation of Mining Activities (2010)	Regulation on the Implementation of Mining Activities (2010)
Regulation on Permission of Mining Activities	2005	-	Regulation on Permission Mining Activities (2005)
Regulation on Reclamation of Lands Disturbed by Mining Activities	2007	Regulation on Reclamation of Lands Disturbed by Mining Activities (2010)	Regulation on Reclamation of Lands Disturbed by Mining Activities (2010)

Regulation on Protection of Groundwater against Pollution and Deterioration	2012	-	Regulation on Protection of Groundwater against Pollution and Deterioration (2012)

CONCLUDING REMARKS

One of the human footprints that cause drastic changes on environment is mining. Although it has a significant contribution to world economy and an indisputable social influence on the life of communities, its devastating negative impacts on environment cannot be disregarded. Particularly, opencast mining activities severely alter the topography and the physical conditions of the atmosphere, and inversely affect plant life, soil conditions, wildlife habitats, and water resources in the mining area and in its immediate surroundings.

As a result of above mentioned factors, post-mining landscapes lose their previous aesthetic, ecological and socioeconomic values. If effective mitigation measures are not taken to decrease the adverse environmental impacts, environmental degradation due to opencast mining operations may be irreversible.

As addressed within the chapter, the ultimate goal of reclamation is two-fold: i) to sustainably establish the aesthetic and ecological conditions of the post-mining landscape so that it become as much compatible as with surrounding undisturbed lands, and ii) to regain or enhance the productive capacity and stability of the land so that it contributes to community's economic and social welfare in a more efficient way.

Due to rapid industrialization and economic growth, the size and the content of the problems arising from negative impacts of mining activities have been changed and become more complicated than ever. So, in order to achieve successful results in reclamation studies, multidisciplinary approach enriched with the latest technological means is highly required. Of course, there is no *"unique"* and *"magical"* reclamation plan that can be directly applied on all post-mining areas, since major determinants in each reclamation study highly differ and depend on the specific characteristics of the site. Additionally, collaborative and creative involvement of all concerned parties (i.e., state and company officials, local authorities and non-governmental organizations, scientist, engineers and specialists, environmental groups etc.) is crucial for the development of permanently stable landscape use and reclamation plans. It is also necessary to emphasize that reclamation studies should begin at the earliest stages of project development, continue during mining, and proceed after the operation is completed.

The role of landscape architects in such studies has recently gone far beyond the "classical" borders of the profession. Instead of routine beautification and site planning tasks, now they often involve in large-scale complex reclamation and rehabilitation projects, and they even serve as the leader of the project team by taking the advantage of their education and practical experiences, which enables them to develop more innovative, consolidative and comprehensive approaches toward the optimum solution.

Legislative issues in mining and reclamation studies are mostly contingent to the visions of the governments. However, in order to foster efficiency and sustainability of post-mining landscapes, and to protect our valuable natural resources, much stricter global standardization on legal measures is needed in our rapidly changing world.

Our future depends on what we do today and how we interact with nature. So, it is essential to sustainably reclaim mine-disturbed lands through comprehensive and collaborative planning that considers all key aspects. Because we borrow the nature that we live in from future generations, which is a fact that we should always recall.

REFERENCES

1.   N. Akpinar, D. Kara, E. Ünal, Post surface mining land use planning. Turkey XIII. Mining Congress, 1014May 1993stanbul: Chamber of Mining Engineers of Turkey, 1993.

2.   N. Akpinar, The process of revegetation in the post-mining reclamation. The Mining and Environment Symposium, 56May 2005Ankara; 2005.

3.   Alcoa. Jarrahdale. http://www.alcoa.com/australia/en/info_page/ mining_jarrahdale.asp (accessed 18.12.2012).

4.   S. T. Anderson, The Mineral Industry of Germany. 2010Minerals Yearbook, Germany [Advance Release], U.S. Department of the Interior U.S. Geological Survey; 2012.

5.   B. Arbogast, Interrogating a Landscape Design Agenda in the Scientifically Based Mining World. In: Berger A. (ed.) Designing the Reclaimed Landscape. London and New York: Taylor&Francis Group; 20085360

6.   A. M. Bauer, Reclemation Planning of Pits and Quarries. Landscape Architecture Technical Information Series (LATIS). Washington, USA: The American Society of Landscape Architects; 2000

7.   F. G. Bell, S. E. T. Bullock, T. F. J. Halbich, P. Lindsay, Environmental Impacts Associated with an Abandoned Mine in the Witbank Coalfield, South AfricaInternational Journal of Coal Geology200145195216

8.   F. G. Bell, L. J. Donnelly, Mining and Its Impact on the EnvironmentOxon, England: Taylor&Francis Group; 2006

9.   G. Betlem, E. Brans, K. Getliffe, F. Groen, Environmental Liability & Mining Law in Europe. Environmental Regulation of Mine Waters in the European Union (ERM TE) WP5 Report. UK; 2002

10.   A. Bhattacharya, J. Routh, G. Jacks, P. Bhattacharya, M. Mörth, Environmental Assessment of Abandoned Mine Tailings in Adak, Vasterbotten District (Northern Sweden). Applied Geochemistry 20062117601780

11.   A. D. Bradshaw, Underlying Principles of Restoration. Can. J. Fish. Aquat. Sci. 199653139

12.   D. A. Bruns, T. O. Sweet, Geospatial tools to support watershed environmental monitoring and reclamation: assessing mining impacts on the upper Susquehanna- Lackawanna American Heritage River. Paper presented at the Office of Surface Mining National Geospatial Conference 200479December 2004, Atlanta, GA.; 2004

13.   D. J. Buchanan, D. Brenkley, Green Cool Mining. In: Hester R.E., Harrison R.M. (ed.) Mining and its Environmental Impact. Cambridge: Royal Society of Chemistry, Thomas Graham House; 19947195

14. X. Cao, Regulating Mine Land Reclamation in Developing Countries: The Case of China. Land Use Policy 200724472483

15. Chamber of Mining Engineers of Turkey. Legislation. Professional Legislation http://www.maden.org.tr/mevzuat/mesleki_mevzuat. php (accessed 10 December 2012).

16. Chamber of Mines of South Africa. CM: Mining and Environmental Impact Guide. Gauteng Department of Agriculture, Environment and Conservation. 2008.http://www.bullion.org.za/Departments/ Environment/Downloads/Impact%20Guide/GDACE%20 Mining%20and%20Environmental%20Impact%20Guide. pdf(accessed 02.12.2012).

17. J. Cherry, Case Study of Successful Reclamation and Sustainable Development at Kennecott Mining Sites. In: Berger A. (ed.) Designing the Reclaimed Landscape. London and New York: Taylor&Francis Group; 2008105112

18. S. Chevrel, V. Kuosmannen, S. Marsh, H. Mollat, P. Vosen, E. Kuronen, Hyperspectral airborne imagery for mapping mining-related contaminated areas in various European environments-first results of the mineo project. Paper presented at the International Airborne Remote Sensing Conference. 1720September 2001, san Francisco, California; 2001

19. S Chevrel, ., R Belocky, ., K Grösel, . Monitoring and Assessing the Environmental Impact of Mining in Europe Using Advanced Earth Observation Techniques- MINEO, First Results of the Alpine Test Site. In: Phillmann W., Tochtermann K. (eds.) Environemental Communication in the Information Society, EnviroInfo Vinee 2002. Part 1, 518526 . http://www2.brgm.fr/mineo/Publications/ Enviroinfo_paper.pdf (accessed 15 December 2012).

20. P. Coelho, S. Silva, J. Roma-torres, C. Costa, A. Henriques, J. Teixeira, M. Gomes, O. Mayan, Health Impact of Living Near an Abandoned Mine-Case Study: Jales Mines. International Journal of Hygiene and Environmental Health 2007210399402

21. W. Cuccu, Site assessment before reclamation. In: Ciccu R. (ed.) SWEMP 2002Proceedings of the 7 th International Symposium on Environmental Issues and Waste Management in Energy and Mineral Production, SWEMP 2002710October 2002, Cagliari, Italy. University of Cagliari; 2002.

22. Davies. Investors' Guide to Mining in Canada. 2011.http://www. dwpv.com/~/media/Files/PDF/Investors-Guide-To-Mining-In-Canada-English.ashx (accessed 12.12.2012).

23. Del Tredici PDisturbance Ecology and Symbiosis in Mine-Reclamation Design. In: Berger A. (ed.) Designing the Reclaimed Landscape. London and New York: Taylor&Francis Group; 20081325

24. E. C. Doll, Relation of Public Policy to Reclamation Goals and Responsibilities. In: Hossner L.R. (ed.) Reclamation of Surface-Mined Lands. Florida: CRC Press, Inc.; 19884153

25. E. M. Fanuscu, The possibilities of evaluation derelict land as urban land in Istanbul Agacli opencast coal mining area. PhD thesis. stanbul University stanbul; 1999

26. Flambeau Reclaimed : Reclamation at the Flambeau Mine : http:// wwwflambeaumine.com/documents/factsheets/reclamation.pdf (accessed 26 October 2012

27. F.D Fox, . Mining and sustainable development Flambeau and Ridgeway mines- lessons learned. 2002. Presented at SME Annual Meeting, 2426 February 2003, Cincinnati, Ohio. http:// www.mining.ubc.ca/mlc/presentations_pub/sme/FoxPaper.pdf (accessed 26 October 2012).

28. J. Frouz, D. Elhottova, V. Kuraz, M. Sourkova, Effects of Soil Macrofauna on Other Soil Biota and Soil Formation in Reclaimed and Unreclaimed Post-mining Sites: Results of A Field Microcosm Experiment. Applied Soil Ecology 200633308320

29. A. Ganas, J. Aerts, T. Astaras, J. De Vente, E. Frogoudakis, N. Lambrinos, C. Riskakis, D. Oikonomidis, A. Filippidis, A. Kassolifournaraki, The Use of Earth Observation and Decision Support Systems in The Restoration Of Open-Cast Nickel Mines in Evia, Central Greece. International Journal of Remote Sensing 2004

30. C. Garavan, J. Breen, R. Moles, O. Regan, B. A Case Study of the Health Impacts in an Abandoned Lead Mining Area, Using Children's Blood Lead Levels. International Journal of Mining, Reclamation and Environment 2008224265284

31. General Directorate of Turkish Coal Enterprises. TKI: Annual report for 2011. http://www.tki.gov.tr/TKI/YillikFaaliyetler/

c1e327f0   -e193-471b-a216-bb07243c3f8bfaal_2011.pdf (accessed 17.12.2012).

32. M. K. Ghose, Air Pollution Due to Opencast Coal Mining and the Characteristics of Air-Borne Dust-an Indian Scenario. International Journal of Environmental Studies 2002592211228

33. H. H. Gillarova, E. Pecharova, An Assessment of the Environmental Impact of the Proposed Medard Lake Project. Journal of Landscape Studies 200923341

34. Y. Gorokhovich, M. Reid, E. Mignone, Prioritizing Abandoned Coal Mine Reclamation Projects within the Contiguous United States Using Geographic Information System Extrapolation. Environmental Management 2003324527534

35. E. Görcelioglu, Landscape Reclamation Techniques. stanbul: Emek Printing; 2002

36. B. Groombridge, Global Biodiversity: Status of the Earth's Living Resources. New York: Chapman & Hall; 1992

37. G. R. Hancock, The Use of Landscape Evolution Models in Mining Rehabilitation Design. Environmental Geology 200446561573

38. G. R. Hancock, M. K. Grabham, P. Martin, K. G. Evans, A. Bollhöfer, A Methodology for the Assessment of Rehabilitation Success of the Post-mining Landscapes-Sediment and Radionuclide Transport at the Former Nabarlek Uranium Mine, Northern Territory, Australia. Science of the Total Environment 2006354103119

39. G. D. Harwood, J. L. Thames, Design and Planning Considerations in Surface-Mined Land Shaping. In: Hossner L.R. (ed.) Reclamation of Surface-Mined Lands. Florida: CRC Press, Inc.; 1988137158

40. M. Hendrychova, Reclamation Success in Post-Mining Landscapes in the Czech Republic: A Review of Pedological and Biological Studies. Journal of Landscape Studies 200816378

41. D. Hladnik, Spatial Structure of Disturbed Landscapes in Slovenia. Ecological Engineering 2005241727

42. M. Hunt, Mining Law in Western Australia (4th Ed.). The Federation Press, 2009

43. R. J. Hutnik, G. W. Mckee, Reclamation (Revegetation). In: Kennedy B.A. (ed.) Surface Mining. 2nd Edition. Littleton, Colorado: Society for Mining, Metallurgy, and Exploration, Inc.; 1990811817

44. R. F. Hüttl, Ecology of Post Strip-Mining Landscapes in Lusatia, Germany. Environmental Science and Policy 19981129135

45. R. F. Hüttl, W. Gerwin, Landscape and Ecosystem Development after Disturbance by Mining. Ecological Engineering 20052413

46. J.M.N Ibarra, ., de las Heras M.M. Open-Cast Mining Reclamation. In: Mansourian S., Vallauri D., Dudley N. (ed.) Forest Restoration in Landscape: Beyond Planting Trees. New York: Springer Verlag; 2005. 370376 . Available from http://www.bf.uni-lj.si/fileadmin/groups/2716/downloads/%C4%8Clanki_vaje/Mansurian_forest_restoration_53_mining.pdf (access 14 October 2012).

47. I. J. Jansen, S. W. Melsted, Land Shaping and Soil Construction. In: Hossner L.R. (ed.) Reclamation of Surface-Mined Lands. Florida: CRC Press, Inc.; 1988125136

48. Y. Jun-bao, L. Jing-shuang, W. Jin-da, L. Zhong-gen, Z. Xue-lin, Spatial-Temporal Variation of Heavy Metal Elements Content in Covering Soil of Reclamation Area in Fushun Coal Mine. Chinese Geographical Science 2002123268272

49. C. Kavourides, F. Pavloudakis, P. Filios, Environmental protection and land reclamation works in West Macedonia Lignite Centre in North Greece current practice and future perspectives. In: Ciccu R. (ed.) SWEMP 2002: Proceedings of the 7 th International Symposium on Environmental Issues and Waste Management in Energy and Mineral Production, SWEMP 20027100ctober 2002, Cagliari, Italy. University of Cagliari; 2002

50. A. J. Kibble, P. J. Saunders, Contaminated Land and the Link with Health. In: Hester R.E., Harrison R.M. (ed.) Assessment and Reclamation of Contaminated Land. Cambridge: Royal Society of Chemistry, Thomas Graham House; 20016584

51. M. Kibblewhite, Identifying and Dealing with Contaminated Land. In: Hester R.E., Harrison R.M. (ed.) Assessment and Reclamation of Contaminated Land. Cambridge: Royal Society of Chemistry, Thomas Graham House; 20014564

52. A. Kleeberg, A. Schapp, D. Biemelt, Phosphorus and Iron Erosion from Non-Vegetated Sites in a Post-Mining Landscape, Lusatia, Germany: Impact on Aborning Mining Lakes. Catena 200872315324

53. A. Klukanova, S. Rapant, Impact of Mining Activities upon the Environment of the Slovak Republic: Two Case Studies. Journal of Geochemical Exploration 199966299306

54. W. Knabe, Methods and Results of Strip-Mine Reclamation in Germany. The Ohio Journal of Science 196464275105

55. K. Komnitsas, K. Modis, Soil Risk Assessment of As and Zn Contamination in A Coal Mining Region Using Geostatisretics. Science of the Total Environment 2006371190196

56. Kutorman. Past to present. http://www.kutorman.com/ (accessed 18.12.2012).

57. M. Kyzeridi, G. N. Panagiotou, S. Peppas, Reclamation scenarios of mined-out open pits using GIS tools. In: Ciccu R. (ed.) SWEMP 2002Proceedings of the 7 th International Symposium on Environmental Issues and Waste Management in Energy and Mineral Production, SWEMP 2002710October 2002, Cagliari, Italy. University of Cagliari; 2002.

58. Legislation.gov.uk. Search all legislation. http://www.legislation.gov.uk/ (accessed 20 December 2012)

59. Z. Miao, R. Marrs, Ecological Restoration and Land Reclamation in Open-Cast Mines in Shanxi Province, China. Journal of Environmental Management 200059205215

60. R. W. Micsak, The Legal Landscape. In: Berger A. (ed.) Designing the Reclaimed Landscape. London and New York: Taylor&Francis Group; 2008154164

61. Mineral Information Institute. Mining Reclamation Success- Coal Mining Reclamation. http://www.mii.org/Rec/coal/coal.html (accessed 18.12.2012).

62. D. L. Mummey, P. D. Stahl, J. S. Buyer, Microbial Biomarkers as an Indicator of Ecosystem Recovery Following Surface Mine Reclamation. Applied Soil Ecology 200221251259

63. A. Navarro, D. Collado, M. Carbonell, J. A. Sanchez, Impact of Mining Activities on Soils in a Semi-Arid Environment: Sierra Almagrera District, SE Spain. Environmental Geochemistry and Health 200426383393

64. M. C. Navarro, C. Perez-sirvent, M. J. Martinez-sanchez, J. Vidal, P. J. Tovar, J. Bech, Abandoned Mine Sites As A Source of Contamination by Heavy Metals: A Case Study in A Semi-Arid Zone. Journal of Geochemical Exploration 200896183193

65. Office of Surface Mining Reclamation and Enforcement. OSM: Surface Mining Law: Public Law 9587 . Surface Mining Control and Reclamation Act of 1977 (SMCRA). http://www.osmre.gov/ topic/smcra/SMCRA.pdf (accessed 13 October 2012).

66. Official Gazette of Republic of Turkey. http://www.resmigazete. gov.tr/default.aspx (accessed 01 December 2012).

67. B. Passariello, V. Giuliano, S. Quaresima, M. Barbara, S. Caroli, G. Forte, G. Carelli, I. Iavicoli, Evaluation of the Environmental Contamination at an Abandoned Mining Site. Microchemical Journal 200273245250

68. E. Pecharova, H. Broumova-dusakova, K. Novotna, I. Svoboda, Function of Vegetation in New Landscape Units After Brown Coal Mining. International Journal of Mining, Reclamation and Environment 2011254367376

69. J. L. Powell, Revegetation Options. In: Hossner L.R. (ed.) Reclamation of Surface-Mined Lands. Florida: CRC Press Inc.; 19884991

70. R. Pöykiö, P. Perämäki, R. Bergstrom, T. Kuokkanen, H. Rönkkömäki, Assessment of the Impact of Opencast Chrome Mining on the Ambient Air Concentrations of TSP, Cr, Ni and Pb Around A Mining Complex in Northern Finland. International Journal of Environmental Analytical Chemistry 2002825307319

71. I. Prikryl, I. Svoboda, P. Sklenicka, Restoration of landscape functions at area devastated by opencast brown coal mining in the Northwest Bohemia. In: Ciccu R. (ed.) SWEMP 2002: Proceedings of the 7 th International Symposium on Environmental Issues and Waste Management in Energy and Mineral Production, SWEMP 2002710October 2002, Cagliari, Italy. University of Cagliari; 2002

72. R. V. Ramani, R. J. Sweigard, M. L. Clar, Reclamation (Reclamation Planning). In: Kennedy B.A. (ed.) Surface Mining. 2nd Edition. Littleton, Colorado: Society for Mining, Metallurgy, and Exploration, Inc.; 1990750769

73. I. Razo, L. Carrizales, J. Castro, F. Diaz-barriga, M. Monroy, Arsenic and Heavy Metal Pollution of Soil, Water and Sediments in a Semi-Arid Climate Mining Area in Mexico. Water, Air, and Soil Pollution 2004152129152

74. Republic of Turkey Ministry of Justice. Legislation. http://www.mevzuat.adalet.gov.tr/ (accessed 01 December 2012).

75. Republic of Turkey Prime Ministry. Legislation Information System. Regulations. http://www.mevzuat.gov.tr/Yonetmelikler.aspx (accessed 01 December 2012).

76. D. L. Rimmer, A. Younger, Land Reclamation after Coal-mining Operations. In: Hester R.E., Harrison R.M. (ed.) Contaminated Land and its Reclamation. Cambridge: Royal Society of Chemistry, Thomas Graham House; 19977390

77. H. B. Sahu, S. Dash, Land Degradation due to Mining in India and its Mitigation Measures. Proceedings of the 2nd International Conference on Environmental Science and Technology, IPCBEE 6February 26-28, 2011Singapore: IACSIT Press; 2011.

78. V. P. Salonen, S. Valpola, N. Tuovinen, Impacts of mining on the environment in Finland. Seattle Annual Meeting. 25November 2003, Seattle, Washington. Paper 199-7Geological Society of America Abstracts with Programs 2003

79. J.I Sams, ., K.M Beer, . Effects of Coal-Mine Drainage on Stream Water Quality in the Allegheny and Monongahela River Basins-Sulfate Transport and Trends: Water-Resources Investigations Report 994208 , Etc., Lemoyne, Pennsylvania; 2000. http://pa.water.usgs.gov/reports/wrir_99-4208.pdf (accessed 15 December 2012).

80. L. W. Saperstein, Reclamation (Introduction). In: Kennedy B.A. (ed.) Surface Mining. 2nd Edition. Littleton, Colorado: Society for Mining, Metallurgy, and Exploration, Inc.; 1990749

81. H. J. Schor, D. H. Gray, Landforming: An Environmental Approach to Hillside Development, Mine Reclamation and Watershed Restoration. Hoboken, New Jersey: John Wiley&Sons, Inc; 2007

82. M. Sengupta, Environmental Impacts of Mining Monitoring, Restoration, and Control. USA: Lewis Publishers; 1993

83. P. Sklenicka, I. Prikryl, I. Svoboda, T. Lhota, Non-Productive Principles of Landscape Rehabilitation after Long-Term Opencast Mining in North-West Bohemia. The Journal of the South African Institute of Mining and Metallurgy 20048388

84. P. Sklenicka, I. Kasparova, Restoration of Visual Values in a Post-Mining Landscape. Journal of Landscape Studies 20081110

85. C. R. Smyth, P. Dearden, Performance Standards and Monitoring Requirements of Surface Coal Mine Reclamation Success in Mountainous Jurisdictions of Western North America: A Review. Journal of Environmental Management 199853209229

86. S. Song, E. S. Min, M. H. Kim, H. K. Lee, Pollution by Acid Mine Drainages from the Daeseong Coal Mine in Keumsan. Econ. Environ. Geol 1997302105116

87. ile Forestry Operation Directorate. Current documents. Presentation 2012.http://www.sileorman.gov.tr/ (accessed 18.12.2012).

88. F. im ir, Ç. Pamukçu, M. K. Özfirat, Mine Reclamation and Restoration of Nature. Dokuz Eylul University Faculty of Engineering Journal of Science and Engineering 2007923949

89. M. Topay, Sertkaya Aydın , Koçan N. Ta Ocaklarının Peyzaja Etkileri ve Yeniden Kullanımlarına Yönelik Çözüm Önerileri: Bartın li Örne i. Süleyman Demirel Üniversitesi Orman Fakültesi Dergisi 2007A(2): 134-144.

90. T. Tören, Determination of environmental impacts due to open pit coal mining activities: a case study from Turkey. In: Ciccu R. (ed.) SWEMP 2002Proceedings of the 7 th International Symposium on Environmental Issues and Waste Management in Energy and Mineral Production, SWEMP 2002710October 2002, Cagliari, Italy. University of Cagliari; 2002.

91. N. A. Tshivhandekano, Documenting reclamation and closure of Ermelo coal mines (Mpumalanga Province): Implications for developing a national strategy for mine reclamation in South Africa. MSc. thesis. Master of Arts (Environment and Society), Department of Geography, Geoinformatics & Meteorology, Faculty of Humanities, University of Pretoria South Africa; 2004

92. P. Wood, Remediation Methods for Contaminated Sites. In: Hester R.E., Harrison R.M. (ed.) Contaminated Land and its Reclamation. Cambridge: Royal Society of Chemistry, Thomas Graham House; 19974771

93. P. Wood, Remediation Methods for Contaminated Sites. In: Hester R.E., Harrison R.M. (ed.) Assessment and Reclamation of Contaminated Land. Cambridge: Royal Society of Chemistry, Thomas Graham House; 2001115139

94.   W Xin-yi, ., Y Jian, ., G Hui-xia, . Research of the Change of Heavy Metal Concentration in the Soil around the Coal Mining Waste Dump. 2009.http://old.cgs.gov.cn/zt_more/34/zhaiyao/html/06/609.htm. (accessed 25.04.2010).

95.   P. L. Younger, Environmental Impacts of Coal Mining and Associated Wastes: A Geochemical Perspective. Geological Society 2004236169209

Geochemical Speciation and Risk Assessment of Heavy Metals in Soils and Sediments

Santosh Kumar Sarkar[1], Paulo J.C. Favas[2, 3],
Dibyendu Rakshit[1], and K.K. Satpathy[4]

[1]Department of Marine Science, University of Calcutta, Calcutta, West Bengal, India

[2]Department of Geology, School of Life Sciences and the Environment, University of Trás os-Montes e Alto Douro, Vila Real, Portugal

[3]IMAR-CMA Marine and Environmental Research Centre, Faculty of Sciences and Technology, University of Coimbra, Coimbra, Portugal

[4]Indira Gandhi Centre for Atomic Research, Environment and Safety Division, Kalpakkam, Tamil Nadu, India

INTRODUCTION

Heavy metal pollution is a serious and widely environmental problem due to the persistent and non-biodegradable properties of these contaminants. Sediments serve as the ultimate sink of heavy metals in the marine environment and they play an important role in the transport and storage of potentially hazardous metals. They are introduced into the aquatic system as a result of weathering of soil and rocks, from volcanic eruptions and from a variety of human activities involving mining, dredging, processing and use of metals and/or substances containing metal contaminants. Heavy metals entering natural water become part of the water-sediment system and their distribution processes are controlled by a dynamic set of physicochemical interactions and equilibria The properties of metals in soils and sediments depend on the physiochemical form in which they occur [1]. Heavy metals are distributed throughout soil and sediment components and associated with them in various ways, including adsorption, ion exchange, precipitation and complexation and so on [2]. Changes in environmental conditions, such as temperature, pH, redox potential and organic ligand concentrations, can cause metals to be released from solid to liquid phase and sometimes cause contamination of surrounding waters in aquatic systems [3]. They are not permanently fixed by soil or sediment. Therefore, it cannot provide sufficient information about mobility, bioavailability and toxicity of metals if their total contents are studied alone.

Natural and anthropogenic activities have the capacity to cause changes in environment conditions, such as acidification, redox potential, or organic ligand concentrations, which can remobilize contaminated soils and sediments releasing the elements from soils and sediments and pore water to the water column resulting contamination of surrounding waters. Daily tidal currents, wind energies, and storms in coastal and estuarine systems can cause periodical remobilization of surface sediments [4]. More turbulent conditions, such as seasonal flooding or storms, or bioturbation, due to feeding and movement of benthic organisms, can expose anoxic sediments to oxidant conditions. In addition, activities such as dredging result in major sediment disturbances, leading to changes in chemical properties of sediment [5].

The remediation of heavy metal pollution is often problematic due to their persistence and non-degradability in the environment. As a sink and source, soils and sediments constitute a reservoir of bioavailable heavy metals and play a significant role in the remobilization of contaminants in the aquatic systems under favorable conditions. Such potential of sediment for being a sink as well as a source of contaminant can make sediment chemistry and toxicity key components of the quality of aquatic system. Much concern has been focused on the investigation of the total element contents in soils and sediments. However, it cannot provide sufficient information about mobility, bioavailability and toxicity of elements and thus may not be able to provide information about the exact dimension of pollution. The data on total contents of metals are quite insufficient to estimate the possible risk of remobilization of total metals under changing environmental conditions and potential uptake of liberated metals by biota and thus the determination of different fractions assume great importance. This has been described as "speciation" [6]. Since each form have different bioavailability and toxicity, the environmentalists are rightly concerned about the exact forms of metal present in the aquatic environment.

The concept of speciation dates back to 1954 when Goldberg introduced the concept of speciation to improve the understanding of the biogeochemical cycling of trace elements in seawater. Kinetic and thermodynamic information together with the analytical data made it possible to differentiate between oxidized versus reduced, complexed or chelated versus free metal ions in solution and dissolved between particulate species. Florence [7] has defined the term speciation analysis as the determination of the individual physicochemical forms of the element, which together make up its total concentration in a sample. According to Lung [8], speciation analysis involves the use of analytical methods that can provide information about the physicochemical forms of the elements. Schroeder [9] distinguishes physical speciation, which involves differentiation of the physical size or the physical properties of the metal, and chemical speciation, which entails differentiation among the various chemical forms. The main objective of measuring metal species relates to their relative toxicities to aquatic biota. The second and long term aim of speciation studies is to advance an understanding of metal interactions between water and bed sediments in an aquatic ecosystem. In the last decade researchers have followed different sequential extraction techniques for the fractionation of metals in sediments of different river systems. Rauret et al. [10] studied the speciation of copper and lead in the sediments of River Tenes (Spain) while Pardo et al. [11] studies the speciation of zinc, cadmium, lead, copper, nickel and cobalt in the sediments of

Pisuerga River, Spain, in order to establish the extent to which these are polluted and their capacity to remobilization. Jardo and Nickless [12] investigated the chemical association of zinc, cadmium, lead and copper in soils and sediments of England and Wales. In most samples, these four metals were associated with all the chemical fractions. Tessier et al. [13] studied speciation of cadmium, cobalt, copper, nickel, lead, zinc, iron and manganese in water and sediments of St. Fransois River, Quebec, Canada. Elsokkary and Muller [14] studied speciation of chromium, nickel, lead and cadmium in the sediments of Nile River, Egypt, reporting that a high proportion of chromium, nickel and lead are bound to organic material and sulphides, while cadmium is bound to carbonate fraction. Ure [15] and Rauret [16] have reviewed the chemical extraction procedures used for heavy metal determinations in contaminated soils and sediments. Owing to the need for validation of extraction schemes, the EC Measurement and Testing Programme (formerly BCR) has organized a project for improving the quality of determinations of extractable heavy metals, where development and validation of extraction procedures has been discussed [17, 18].

The present article aims to summarize the potentials of sequential extraction technique adopting different analytical protocols for gaining information on the mobility and dynamics of operationally determined chemical forms of heavy metals in soils and sediments. The BCR (Community Bureau of Reference, now superseded by the Standards, Measurement and Testing Programme of the European Community) procedure has been illustrated considering the case study of Ganges (Hugli) River Estuary and adjacent Indian Sundarban mangrove wetland (a UNESCO World Heritage Site), northeastern part of the Bay of Bengal. In addition, the authors also evaluate the modified BCR sequential extraction technique as devised by various scientists, the risk assessment code (RAC) as well as assessment of toxicity comparing with sediment quality guidelines. The RAC classification is based on the strength of the bond between the metals and the different geochemical fractions in sediments or soils and the ability of metals to be released and enter into the food chain.

SEQUENTIAL EXTRACTION: MERITS AND DEMERITS

The sequential extraction provides more or less detailed information concerning the origin, mode of occurrence, biological and physicochemical availabilities, mobilization and transport of heavy metals. The procedure stimulates the mobilization and retention of these species in the natural environment using changes in environmental condition such as pH, redox potential and degradation of organic matter [16]. A series of reagents is applied to the sample, increasing the strength of the extraction at each step, in order to dissolve the trace metal present in different sediment phases. The extractants are inert electrolytes, weak acids, reducing agent, oxidizing agents and strong mineral acids [19].

The 3-stage sequential extraction procedure proposed by the European Community Bureau of Reference (BCR) was developed in an attempt to standardize the various schemes described in the literature [2, 20, 21], since the use of different procedures, varying in the number of steps, types of reagents and extraction condition. Hindered comparison of results obtained in the many studies of heavy metals chemical fractionation in environmental samples [22].

The BCR methods has been widely adapted by various authors, and applied to a range of type of solid sample including fresh water sediment [23-25], salt water sediment [26-28], sewage sludge and particulate matter [29-31]. This scheme enables us to associate the meals with one of the following four geochemical phases:

- *Acid-soluble phase:* This phase is made up of exchangeable metals and others bound to carbonates that are able to pass easily into the water column, for example, when the pH drops. It is the fraction with the most labile bond to the soil/sediment and, therefore, the most dangerous for the environment.

- *Reducible phase:* This phase consists of metals bound to iron and manganese oxides that can be released if the sediment changes from the oxic to the anoxic state, which could be caused, for example, by the activity of microorganisms present in the soils/sediments.

- *Oxidisable phase:* This shows the amount of metal bound to organic matter and sulphides, which can be released under oxidizing conditions. Such conditions can occur, for example, if the sediment is resuspended (by dredging, currents, flooding, tides, etc.) and the sediment participles come into contact with oxygen-rich water.
- *Residual phase:* Lithogenous and inert (Non-bioavailable).

The heavy metals in the soils and sediments are bound to different fractions with different strengths, the value can; therefore, give a clear indication of soil and sediment reactivity, which in turn assess the risk connected with the presence of heavy metals in a terrestrial or aquatic environment. The rationale of the sequential extraction procedure is that each successive reagent dissolves a different component, which can content heavy metals within their crystalline structures. Under natural conditions, metals in minerals are unlikely to experience significant release over the time frames of interest [32, 33].

ANALYTICAL PROTOCOLS FOR SEQUENTIAL EXTRACTION

In recent years a great number of papers have been published on various analytical techniques proposed for the fractionation analysis of trace elements in various environmental samples (soils, sediments, etc.). An approach that has been found to be preferable is the fractionation of heavy metal into operationally defined forms under the sequential action of different extractants [2]. Selective extractants, used in sequential extraction procedures, are aimed at the simulation of natural conditions whereby metals associated with certain soil (sediment) components can be released. For example, changes in the ionic composition affecting adsorption–desorption reactions or a decrease in pH may lead to the release of metals, retained on a matrix by weak electrostatic interactions or co-precipitated with carbonates ("exchangeable" and "acid soluble" forms). Decreasing the redox potential can result in dissolution of oxides, unstable under reducing conditions, and liberation of scavenged metals ("reducible" forms). Changes in oxidizing conditions may cause the degradation of organic matter and release of complexed metals ("oxidizable" forms). Finally, the destruction of primary and secondary mineral lattice releases heavy metal retained within the crystal structure, e.g., due to isomorphous substitution ("residual" forms) [2]. The nominal "forms" determined by operational fractionation can help to estimate the amounts of total metals in different reservoirs which could be mobilized under changes in the chemical properties of the soil [34]. Since the 1970s a considerable number of extraction procedures have been proposed for determining the forms of heavy metal [2, 35-39]. Most of these procedures are based on the scheme of Tessier et al. [2]. Although most of the extracting reagents were originally used in the chemical analysis of soils, the procedures proposed have been tested on a wide variety of contaminated environmental samples—sediments, road dust, sewage sludge, etc.

Sequential extraction can be useful to have an operational classification of metals in different geochemical fractions [2] which is the most reliable criteria to quantify the potential effect of soil/sediment contamination by heavy metals. This can provide information about the identification of the main binding sites, the strength of element binding to the particulates and the phase associations of trace elements in soil/sediment. Following this basic scheme, some modified procedures with different sequences of reagents or operational conditions have been developed [40-43]. Considering the diversity of procedures and the lack of uniformity in different protocols, a European Community Bureau of Reference (BCR, now the European Community Standards Measurement and Testing Program) method was proposed [6] and was applied by a large group of researchers [31, 44-47]. In this study, we followed the sequential extraction procedure proposed by the European Union›s Standards, Measurements and Testing program [3].

MODIFIED BCR SEQUENTIAL EXTRACTION PROCESS

As discussed above it is evident that sequential extraction provides valuable information regarding identification of main binding site, the strength of the element binding to the particulates and the phase associations of heavy metals in sediments. However, various complicated sequential extraction procedures were experimented to provide more detailed information regarding different metal phase associations [2, 48, and 49]. A wide range of techniques is available whereby various extraction reagents and experimental conditions are used. These techniques involve a 5-step [2], 4-step (BCR, Bureau Commune de Reference of the European Commission), 6-step [50] and 7-step [51, 52] extraction, and are thus becoming popular methods to be used for sequential extraction [53, 54]. Following this basic scheme, some modified procedures with different sequences of reagents or operational conditions have been developed [40-43].

Several sophisticated instruments have been used for the determination of heavy metals contents in marine environments. These include; flame AAS [55, 56], atomic fluorescence spectrometry [57], anodic stripping voltametry [58, 59], ICP-AES [60] and ICP-MS [61, 62].

Heavy metal mobility and bioavailability depend strongly on their chemical and mineralogical forms in which they occur [63]. Several speciation studies have been conducted to determine study different forms of heavy metals rather their total metal content. These studies reveal the level of bioavailability of metals in harbour sediments and also confirm that sediments are indicators of heavy metal pollution in marine environment [64-67].

Since the early 1980s and 1990s sequential extraction methodology has been developed to determine speciation of metals in sediments [2, 68] due to the fact that the total concentration of metals often does not accurately represent their characteristics and toxicity. In order to overcome the above mentioned obstacles it is helpful to evaluate the individual fractions of the metals to fully understand their actual and potential environmental effects [2]. To date, strong acid digestion is used often for the determination of total heavy metals in the sediments. However, this method can be misleading when assessing environmental effects due to the potential for an overestimation of exposure risk. Moreover, in order to determine the mobility of heavy metals in sediments, various sequential extraction procedures have been developed [69-71].

Among a range of available techniques using various extraction reagents and experimental conditions to investigate the distribution of heavy metals in sediments and soils, the 5-step Tessier et al. [2] and the 6-step extraction method, Kersten and Fronstier [50] were mostly widely used. Following these two basic schemes, some modified procedures with different sequences of reagents or experimental conditions have been developed [40-43]. Considering the diversity of procedures and lack of uniformity in different protocols, a BCR, Bureau Commun de Recherche (now called the European Community (EC) Standards Measurement and Testing Programme) method was proposed [6]. It harmonized differential extraction schemes for sediment analysis. The method has been validated using a sediment certified reference material BCR-701 with certified and indicative extractable concentration of Cd, Cr, Cu, Ni, Pb and Zn [72]. This method was applied and accepted by a large group of specialists [31, 44, 45, 47, 73, and 74] despite some shortcoming in the sequential extraction steps [75, 76].

Wang et al. [77] used a modified Tessier sequential extraction method to investigate the distribution and speciation of Cd, Cu, Pb, Fe, and Mn in the shallow sediments of Jinzhou Bay, Northeast China. This site was heavily contaminated by nonferrous smelting activities. They found out that the concentrations of Cd, Cu and Pb in sediments was to be 100, 73, 13 and 7 times, respectively, and higher than the National guidelines (GB 18668-2002). The sequential extraction tests revealed that 39%-61% of Cd was found in exchangeable fractions. This shows that Cd in the sediments posed a high risk to the local environment. Copper and Pb were found to be at moderate risk levels. According to the relationships between percentage of metal speciation and total metal concentration, it was concluded that the distributions of Cd, Cu and Pb in some geochemical fractions were dynamic in the process of pollutants migration and stability of metals in marine sediments from Jinzhor Bay decrease in the order Pb>Cu>Cd.

Yuan et al. [78] applied BCR-sequential extraction protocol to obtain metal distribution patterns in marine sediments from the East China Sea. The results showed that both the total contents and the most dangerous non-residual fractions of Cd and Pb were extremely high. More than 90% of the total concentration of V, Cr, Mo and Sn existed in the residual fraction while more than 60% of Fe, Co, Ni, Cu, and Zn were mainly present in the residual fraction. Manganese, Pb, and Cd were dominantly present in the non-residual fractions in the top sediments.

Jones and Turki [79] worked on distribution and speciation of heavy metals in surface sediments from the Tees estuary, North East England. Tessier et al. [2] metal speciation scheme modified by Ajay and van Loon

[80] was used for the study. They observed out that the sediments were largely organic-rich clayey silts in which metal concentrations exceed background levels, and which attain peak values in the upper and middle reaches of the estuary. Chromium, Pb and Zn were associated with the reducible, residual, and oxidizable fractions. Cobalt and Ni were not highly enriched while Cu is associated with the oxidizable and residual fractions. Cadmium is associated with the exchangeable fractions.

Pempkowlak et al. [81] investigated the speciation of heavy metals in sediments and their bioaccumulation by mussels. They used a 4-step sequential extraction procedure adapted from Forstner and Watmann [82]. Their investigation which was characterized by varying metal bioavailability was aimed at revealing differences in the accumulation pattern of heavy metals in mussel inhabiting that inhabit in sediments. The bioavailabilities of metals were measured using the contents of metals adsorbed to sediments and associated with Fe and Mn hydroxides. The biovailable fraction of heavy metals contents in sediments collected from Spitsbergen represented a small proportion (0.37% adsorbed metals and 0.11%, are associated with metals hydroxides). It was also revealed that the percentages of metals adsorbed and bound to hydroxides of the sediments ranged from 1 to 46% and 1 to 13%, respectively.

Wepener and Vermeulen [66] investigated on the concentration and bioavailability of selected metals in sediments of Richards Bay harbor, South Africa. Sequential extraction of sediments was carried out according to Tessier et al. [2] method. The following metals were investigated: Al, Cr, Fe, Mn, and Zn. Their studies revealed that metals concentrations in sediments samples varied only slightly between seasons, but showed significant spatial variation, which was significantly correlated to sediment particle size composition. Highest metal concentration was recorded in sites with substrates dominated by fine mud. Manganese and Zn had more than 50% of this concentration in reducible fraction while more than 70% of the Cr was associated with the inert fractions and the concentration recorded at some sites were still above action levels when considering only the bioavailable fractions. They also concluded that the concentration of Zn recorded was not elevated their results were compared with the historic data.

Coung and Obbard [54] used a modified 3-step sequential extraction procedure to investigate metal speciation in coastal marine sediments from Singapore as described by the European Community Bureau of Reference (ECBR). Highest percentages of Cr, Ni, and Pb were found in residual fractions in both Kranji (78.9%, 54.7% and 55.9% respectively) and Pulang Tokong (82.8%, 77.3% and 62.2% respectively). This means that these metals were strongly bound to sediments. In sediments from Kranji, the mobility order of heavy metals studied were Cd>Ni>Zn>Cu>Pb>Cr while sediments from Pulan Tekong showed the same order for Cd, Ni, Pb and Cr, but had a reverse order for Cu and Zn (Cu>Zn). The sum of the 4-steps (acid soluble + reducible + oxidizable + residual) was in good agreement with the total metal content, which confirmed the accuracy of the microwave extraction procedure in conjunction with the GFASS analytical method.

Fedotov et al. [83] applied a modified technique for accelerated fractionation of heavy metals in contaminated soils and sediments using rotating coiled columns. Rotating coiled columns (RCC) is valuable for the continuous-flow sequential extraction and can be successfully applied to the dynamic leaching of heavy metals from soil and sediments. This is a fluoroplastic or steel coil wound around a rigid cylindrical drum, which revolves about its axis and, at the same time, revolves around the central axis of the device called planet centrifuge. The stationary (liquid, solid, or heterogeneous) phase is retained in the column because of the centrifugal force field, and the mobile liquid phase is continuously pumped through the column. A solid sample was retrieved in the rotating column as the stationary phase under the action of centrifugal forces while different elements (aqueous solution of complexing reagents, mineral salts and acids) were continuously pumped through. This procedure developed is time saving and requires only 4-5 hr. instead of the several days needed for individual sequential extraction. Losses of solid sample are minimal. Further studies are needed to better estimate the reproducibility of the technique.

Nemati et al. [84] used a modified BCR sequential extraction procedure (SEP) in combination with ICP-MS to obtain the metal distribution patterns in different depths of sediments from Sungai Buloh, Selangor, Malaysia. The results showed that heavy metal contaminations at Sungai Buloh River sediments were more severe than at other sampling sites, especially for Zn, Cu, Ni and Pb. Nevertheless, the element concentrations from top to bottom layers decreased predominantly.

Mossop et al. [85] compared of original and modified BCR sequential extraction procedures for the fractionation of Cu, Fe, Pb, Mn and Zn in soils and sediments. The procedures were applied to five soil and sediment substrates: a sewage sludge-amended soil, two different industrially contaminated soils, river sediment and intertidal sediment. Extractable Fe and Mn concentrations were measured to assess the effects of the procedural modifications on dissolution of the reducible matrix components. Statistical analyses (two-tailed t-tests at 95% confidence interval) indicated that recovery of Fe in step 2 was not markedly enhanced when the intermediate protocol was used. However, significantly greater amounts were isolated with the revised BCR scheme than with the original procedure. Copper behaved similarly to Fe. Lead recoveries were increased by use of both modified protocols, with the greatest effect occurring for the revised BCR extraction. In contrast, Mn and Zn extraction did not vary markedly between procedures. The work indicates that the revised BCR sequential extraction proves better attack on the Fe-based components of the reducible matrix for a wide range of soils and sediments.

SEQUENTIAL EXTRACTION OF METALS IN SEDIMENTS OF THE HUGLI RIVER ESTUARY AND INDIAN SUNDARBAN WETLAND: A CASE STUDY

Materials and Methods

Sample Collection and Sediment Quality Analysis

The delta region formed by Hugli (Ganges) River Estuary (HRE) and is famous for its luxuriant mangrove vegetation, known as Sundarban wetland, acclaimed as UNESCO World Heritage Site for its capacity of sustaining an excellent biodiversity. The wetland is characterized by a complex network of tidal creeks, which surrounds hundreds of tidal islands exposed to different elevations at high and low semi-diurnal tides. This is one of the most sensitive and vulnerable ecosystems in the world and suffers from environmental degradation due to rapid human settlement, tourism and port activities, operation of mechanized boats, deforestation, and increasing agricultural and aquaculture practices. The ongoing degradation is also related to huge siltation, flooding, storm runoff, atmospheric deposition, and other stresses resulting in changes in water quality, depletion of fishery resources, choking of river mouth and inlets, and overall loss of biodiversity. Moreover, the rapid economic development in this deltaic region has caused highly dense areas of human activity and led to serious contamination including heavy metals and persistent organic pollutants (POPs).

Nine sampling sites, namely Barrackpur (S_1), Dakshineswar (S_2), Babughat (S_3), Budge budge (S_4), Ulubaria (S_5), Diamond Harbor (S_6), Frezergunge (S_7), Gangasagar (S_8), and Haribhanga (S_9) were selected considering the existence of typical sediment dispersal patterns along the drainage network systems (as shown in Figure 1) and their position was fixed by a global positioning system (GPS). The stations are representative of the variable environmental and energy regimes that cover a wide range of substrate behavior, wave–tide climate, and intensity of bioturbation (animal–sediment interaction), geomorphological–hydrodynamic regimes and distances from the sea (Bay of Bengal). The sites are exposed to a variable level of heavy metal contamination mainly from anthropogenic sources as mentioned earlier. Six sampling sites (S_1 to S_6) have been chosen along the lower stretch of Hugli River Estuary, while residual three sites (S_7 to S_9) were taken into account in the coastal regions of Sundarban wetland. All sampling sites together with the main stresses to which they are subjected are presented in Table 1.

During winter months (January–March 2009) surface sediment samples weighting 10 g were randomly collected in triplicate from the top 3–5 cm of the surface at each sampling site during low tide using a grab sampler, pooled and thoroughly mixed. Immediately after collection, the samples were placed in sterilized plastic bags in the ice box and transported to the laboratory. Samples were oven dried at 50°C, most gently disaggregated, transferred into precleaned inert polypropylene bags and stored in deep freeze prior to analyses.

Each sample was divided into two aliquots: one unsieved (for the determination of sediment quality parameters) and the other sieved through 63 μm metallic sieves (for elemental analyses). Organic carbon content was determined following a rapid titration method [86] and pH with the help of a deluxe pH meter (model no. 101E) using combination glass electrode manufactured by M.S. Electronics Pvt. Ltd. (India). Mechanical analyses of sediment were done by sieving in a Ro-Tap Shaker manufactured by W.S. Tyler Company, Cleveland, Ohio.

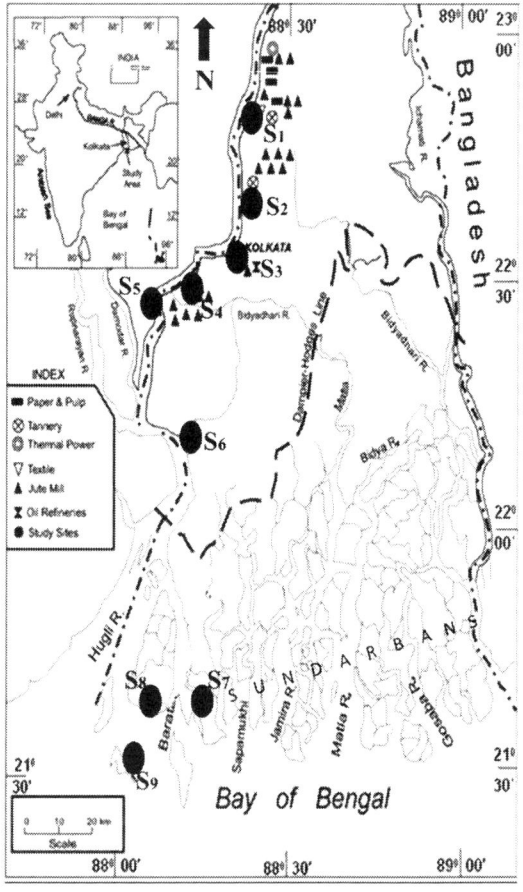

Figure 1: Map showing the location of the monitoring sites (S_1 to S_9) covering Hugli River Estuary and Sundarban mangrove wetland along with the location of the major industries.

Table 1: Details of the nine sampling sites and the main stresses to which they are subjected

Station number	Site	Main stresses
$_s1$	Barrackpur	Industrial effluents, domestic sewage disposal, boating, bathing.
$_s2$	Dakshineswar	Industrial and domestic effluents, boating, bathing, idol immersion site.
$_s3$	Babughat	Power plant discharges, domestic sewage, boating, idol immersion site.
$_s4$	Budge budge	Domestic and industrial effluents, bathing, boating.
$_s5$	Ulubaria	Domestic and industrial effluents
$_s6$	Diamond Harbour	Boating, recreational activities, bathing, fishing, jetties for fishing trawlers
$_s7$	Frezergunge	Tourist activities, ferry services, fishing
$_s8$	Gangasagar	Boating, tourist activities, dredging, fishing, agricultural, domestic and aquaculture practices
$_s9$	Haribhanga	Boating, fishing and ferrying

Analytical Procedure

To determinate the total element concentration, sediment samples were digested in polytetrafluoroethylene vessels with aqua regia (HCl/HNO$_3$, 3:1) and HF neutralized with H$_3$BO$_3$ in a 650 W microwave oven (CEM MDS 2000) with a program consisting of a 20-min ramp and a 30-min hold at 100% power in pressure and temperature controlled conditions (150 psi and 180°C). The digested samples were filtered, transferred to polyethylene containers and stored at +4°C until analysis. All reagents were Suprapur® grade (Merck). Reagent blank was processed with the samples and did not show any significant contamination. Accuracy of the procedure was checked using two different certified reference materials (CRM): MESS-2 and PACS-2, which are both marine sediments certified by the National Research Council of Canada for the element content. The MESS-2 recovery ranged between 91% and 116% for all the elements (Table 2). Precision, calculated as relative standard deviation (RSD %), resulted always lower than 5%.

Table 2: Results of certified reference materials MESS-2 and PACS-2 as well as the observed values All the values are expressed in µg/g of dry weight. MESS-2 and PACS-2 recovery rates are also reported

	Al	As	Cd	Co	Cr	Cu	Fe	Mn	Ni	Pb	Zn
Found MESS-2	86,613±17,773	24.0±2.4	0.230±0.010	15.6±1.5	112±12	38.4±6.1	47,385±3,668	372±42	54.1±3.8	20.0±1.4	170±12
Found PACS-2	70,190±3,784	29.2±1.3	1.59±0.80	12.8±0.6	94.9±4.4	307±22	46,630±1,411	465±23	44.3±2.5	184±10	398±16
Certified MESS-2	85,698±2,600	20.7±0.8	0.240±0.010	13.8±1.4	106±8	39.3±2.0	43,504±2,266	365±21	49.3±1.8	21.9±1.2	172±16
Certified PACS-2	66,125±3,184	26.2±1.5	2.11±0.15	11.5±0.3	90.7±4.6	310±12	43,738±585	440±19	39.5±2.3	183±8	364±23
Recovery MESS-2	101%	116%	95.8%	113%	106%	97.8%	109%	102%	110%	91.2%	99.0%
Recovery PACS-2	106%	111%	75.4%	111%	105%	99.1%	107%	106%	112%	100%	109%

In this study, we followed the sequential extraction procedure proposed by the European Union's Standards, Measurements and Testing program [3]. Selective extraction is based on the procedure used by Tessier et al. [2] with improvements made according to the BCR, which examined and finally eliminated irreproducibility sources. It is made up of three steps, which dissolve the following phases, respectively: exchangeable and bound to carbonate, bound to Fe and Mn oxides and hydroxides, bound to organic matter and sulphides. Exchangeable and bound to carbonate phase (phase 1) is extracted with 0.11 M acetic acid, while the fraction bound to Fe–Mn oxides (phase 2) with 0.5 M hydroxylamine hydrochloride, adjusted to pH 2 with nitric acid (65%). The phase bound to organic and sulphides (phase 3) is extracted with 8.8 M hydrogen peroxide (stabilized at a pH included between 2 and 3), treated at 80°C in a microwave oven using a program consisting of a 30-min ramp and a 60-min hold at 50% power in pressure and temperature controlled conditions (80 psi and 85°C), and 2 M ammonium acetate adjusted to pH 2 with nitric acid (65%). Each extraction was carried out overnight (16 h) at room temperature. All the reagents employed were Tracepur® grade (Merck Eurolab, Italy). After each extraction, the samples were separated from the aqueous phase by centrifuging at 4, 000 rpm for 15 min. The sediments were washed with Milli-Q water and centrifuged again. The wash water was added to supernatants. The element content of the residual phase was obtained from the difference between the total content and the sum of phases 1, 2 and 3, according to Ianni et al. [37, 38], Ramirez et al. [39], and Mester et al. [27]. Sequential extraction reagent blanks showed no detectable contamination. Accuracy of the procedure was checked with BCR-701 (SM&T). The recovery rates for trace elements in the standard reference material ranged between 77% and 118% (Table 3). Precision, calculated as RSD%, resulted generally lower than 5%, except As and Cr in the phase 1 (~20%).

Table 3: Results of certified reference materials BCR-701 as well as the observed values (expressed in µg/g of dry weight) together with recovery rates for each step. n.a. = not available

	Al	As	Cd	Co	Cr	Cu	Fe	Mn	Ni	Pb	Zn
Found BCR-701 step 1	198±1	2.57±0.28	6.09±0.09	2.06±0.08	2.41±0.51	47.7±1.7	43.8±5.8	180±1	14.5±0.3	3.38±0.35	185±4
Found BCR-701 step 2	3,451±46	16.5±0.3	3.37±0.08	3.22±0.03	39.2±0.4	100±2	7,042±106	128±3	24.5±0.4	111±2	102±1
Found BCR-701 step 3	1,912±74	3.09±0.20	0.28±0.01	1.86±0.17	169±4	64.8±1.5	1,147±56	31.9±2.6	17.4±1.7	7.15±0.12	58.4±5.0
Certified BCR-701 step 1	n.a.	n.a.	7.34±0.35	n.a.	2.26±0.16	49.3±1.7	n.a.	n.a.	15.4±0.9	3.18±0.21	205±6
Certified BCR-701 step 2	n.a.	n.a.	3.77±0.28	n.a.	45.7±2	124±3	n.a.	n.a.	26.6±1.3	126±3	114±5
Certified BCR-701 step 3	n.a.	n.a.	0.27±0.06	n.a.	143±7	55.2±4.0	n.a.	n.a.	15.3±0.9	9.3±2.0	54.2±2.0
Recovery step 1	n.a.	n.a.	83.0%	n.a.	107%	96.8%	n.a.	n.a.	94.4%	106%	90.3%
Recovery step 2	n.a.	n.a.	89.4%	n.a.	85.9%	80.6%	n.a.	n.a.	92.3%	87.9%	89.4%
Recovery step 3	n.a.	n.a.	104%	n.a.	118%	117%	n.a.	n.a.	114%	76.9%	108%

The elemental concentrations were determined with an inductively coupled plasma atomic emission spectrometer Vista Pro (Varian), with the internal standard method. Cadmium was determined by electrothermal atomization atomic absorption spectrometry. A Varian Spectra A300 spectrometer with Zeeman Effect background correction and autosampler Varian Model 96 was used employing the standard addition method for calibration. All the metal analyses were performed at the Department of Chemistry and Industrial Chemistry of the University of Genoa (Genoa, Italy).

Statistical Analyses

Principal component analysis (PCA) was used to characterize the metal composition in sediments, and cluster analysis was used for grouping the sampling stations. Principal component analysis (PCA) is a multivariate statistical technique used for data reduction and for deciphering patterns within large sets of data. With PCA, a large data matrix is reduced to two smaller ones that consist of principal component (PC) scores and loadings. PC loadings are eigenvectors of the correlation or covariance matrix depending on which is used for the analysis. The PC scores contain information on all of the variables combined into a single number, with the loadings indicating the relative contribution of each variable to that score [87]. Hierarchical cluster analysis (HCA) characterizes similarities among samples by examining interpoint distances representing all possible sample pairs in high-dimensional space. The sample similarities are represented on two dimensional diagrams call dendrograms [88]. All statistical analyses were performed using the computer software STATISTICA (StatSoft, Inc. 2001).

Results and Discussion

Sediment Geochemistry

Table 4 shows values of pH; organic carbon (%); and percentage of sand, silt, and clay in sediments of the nine sampling sites. Organic carbon values, ranging from 0.22% (in station S_8) to 1.02% (in station S_2), are low in comparison with values found in sediments from other

Indian coastal areas, such as Gulf of Mannar [89], Cochin [90], and Muthupet mangroves [91]. The low organic carbon values might be related with the poor absorbability of organics on negatively charged quartz grains, which predominate in sediments in this estuarine environment [92]. In addition, the constant flushing activity by tides along with the impact of waves can support the low percentage of organic carbon in the sediments. The sediments of the studied stations are characterized by slightly basic pH (7.50–8.36) with maximum values recorded in the stations closest to the sea (stations S_6, S_8, and S_9) and minimum in station S_7.

Table 4: Geographical position, physicochemical and textural properties of sediment samples of 9 sampling sites

Stations	Latitude and Longitude	Salinity	pH	Organiccarbon (%)	Sand (%)	Silt (%)	Clay (%)
S1	22°43' 16" N 88°21' 20 E	0	7.86	0.35	4	87.1	8.9
S2	22°39' 17" N 88°12' 25 E	0	7.80	1.02	1	76.5	22.5
S3	22°33' 53" N 88°20' 19 E	0	7.90	0.52	2.24	41.97	55.79
S4	22°30' 10" N 88°11' 48 E	0–2.5	7.60	0.74	18.25	47.42	34.33
S5	22°28' 06" N 88°06' 54 E	0–1	7.90	0.91	16.7	69.6	13.7
S6	22°11' 14" N 88°11' 15 E	0–5.6	8.36	0.56	3.15	41.13	55.71
S7	21°34' 44" N 88°15' 03 E	30–34.3	7.50	0.36	98.02	0.18	0
S8	21°38' 15" N 88°03' 53 E	32–35	8.14	0.22	32.85	58.45	8.7
S9	21°34' 20" N 88°01' 25 E	35	8.10	0.46	39.3	44.25	16.45

These were different from the low pH values in most of the mangrove swamps in Hong Kong [93], where sediments were not frequently flooded by the tide and become acidic in reducible conditions. With respect to texture, the sediment samples show a variable admixture of sand, silt, and clay. Clay fractions dominate in low-energy areas of suspensional deposits. On the contrary, silt, and sand dominates where the energy level is high. Sediments from station S_7 contain higher percentage of sand (98%) compared to the others, while sediments from S_1, S_2, S_5, and S_8 contains higher percentage of silt (more than 50%) compared to the others. A variable mixture of sand, silt, and clay is present in the other stations and reflect a variable amount of erosion and deposition.

Total Element Concentrations

Total element concentrations in the investigated stations varied in a narrow range of values (Table 5) and were comparable with data obtained for other Indian coastal areas [94, 95]. Datta and Subramanian [96] found very similar trace element concentrations throughout the Bengal Basin, where anthropogenic perturbation is low and river channel may receive a several centimeter-thick sediment layer in a single event during peak flow, preventing to bear the signature of an accumulation of trace elements. The highest concentrations for As, Cu, Fe, Mn, and Ni were measured at station S_9 while for Cd and Pb at station S_3, close to Calcutta city (about 4.5 million residents, but about 14.2 million including suburbs). An anthropogenic input from vehicular traffic and in-dustrial activities may cause high Cd and Pb con-centrations measured in samples collected in the Calcutta urban area. The lowest element concentrations were found at station S_5. Very low (close to the detection limit) Cd concentration was found in the coastal stations (S_7, S_8, and S_9).

Table 5: Total element concentrations (μg/g) in sediments of 9 sampling sites (instrumental precision, calculated as RSD%, resulted lower than 5% for each element in all samples)

Stations	Al	As	Cd	Co	Cr	Cu	Fe	Mn	Ni	Pb	Zn
S1	70,289	8.81	0.165	13.0	67.6	27.8	37,737	591	31.9	20.4	86.6
S2	70,879	8.44	0.452	14.0	74.8	36.8	39,405	625	34.2	22.3	90.7
S3	72,134	8.65	1.79	14.5	73.5	32.3	40,070	712	35.0	33.2	83.1
S4	72,613	8.49	0.492	14.9	76.8	27.9	40,303	726	35.0	19.6	80.4
S5	62,044	6.41	0.220	12.1	58.2	21.1	33,428	597	27.5	17.0	64.1
S6	64,325	6.79	0.106	12.0	64.8	32.0	34,273	613	31.3	17.9	69.6
S7	77,529	7.77	0.044	14.0	75.1	28.4	40,084	389	38.1	20.5	74.4
S8	68,146	8.08	0.027	12.7	62.5	22.2	36,786	511	34.3	19.7	61.4
S9	72,666	9.40	0.044	14.1	74.2	36.6	40,838	785	40.1	22.9	74.9

The geoaccumulation index (I_{geo}) of Muller [97] has been calculated for the analyzed elements, by comparing current concentrations with pre-industrial levels, in order to estimate the metal contamination in sediments. The equation used for the calculation of I_{geo} is: $\log_2 (C_n/1.5 B_n)$, where C_n is the measured content of element "n" and B_n the element's content in "average shale" [98]. Factor 1.5 is used because of possible variations in background values for a given element in the environment, as well as very small anthropogenic influences [99]. As shown in Figure 2, all sediments fall in class 0 for Al, Co, Cr, Cu, Fe, Mn, Ni, Pb, and Zn, therefore the area is not contaminated for these elements. Unlike the Hugli River, in other rivers of the Bengal Basin, such as Meghna and Brahmaputra, Cr exhibits higher I_{geo} values respect to the other elements [96]. For Cd, two stations fall in class 1 and three in class 2 exhibiting a moderate contamination for this element. In all stations, as falls in class 2 (moderate pollution) In this area, as contamination was already observed in previous studies and it is probably due to groundwater contamination [100] This contamination can have natural origin, such as coal seams in Rajmahal basin and arsenic mineral in mineral rocks in the upper reaches of the Ganges river system. The highly reducing nature of groundwater would reduce As, causing the possible desorption of as [101].

Speciation Patterns

The potential environmental risk of trace elements in sediments is associated with both their total content and their speciation. The chemical partitioning of the considered elements (Al, As, Cd, Co, Cr, Cu, Fe, Mn, Ni, Pb, and Zn) from each extraction step has been described. Aluminum, Cr, and Fe are present mainly in the residual phase, representing 95.8–96.8%, 88.9–91%, and 83.0–94.7% of the total concentration, respectively, which implies that these elements are strongly linked to the inert fraction of the sediments. This result was in good agreement with data reported by several studies carried out worldwide in marine coastal areas [45, 46, 78, and 102]. The high percentage of Fe in the residual phase indicates that most of the Fe exists as crystalline Fe peroxides (goethite, limonite, magnetite, etc.). The remaining Fe is associated with the reducible phase (mean, 11.25%). Large amounts of Fe accumulate in the residual phase probably because it is basically of natural origin (it is the most common element in the earth's crust).

Concentrations of Al, Fe, and Cr are very low in exchangeable phase (0.08%, 0.26%, and 1.72% as mean values, respectively), limiting their potential toxicity as pollutants. It should be noted that sediments always act as reservoir for elements; therefore, their potential risk of pollution to environment has always to be considered.

Arsenic, Co, Ni, and to some extent Zn, are found mainly in the residue (~50% of the total concentration). Nickel and Co are associated to the residue respectively for 56% and 74% of the total concentration, with a speciation similar in all the samples. A mean of 23% of Co is present in the phase 2. The highest percentage of labile Co (~13%) was measured in S_6 (Diamond Harbour) and S_8 (Gangasagar) and can be due to a recent input of this element. The dominant proportion of Ni in the residual phase is in agreement with the results of other studies [27, 46]. Nickel is present, apart from the residue, in phases 2 and 3 (about 10% in each phase). Arsenic is distributed mainly between the residual (mean 47%), the reducible, and the oxidizable phases (mean 19% and 22%, respectively). Acharyya et al. [101] observed that as is adsorbed to iron-hydroxide-coated sand grains and to clay minerals in the sediments of the Ganges delta from West Bengal. Among the studied elements, as is found with the greatest proportion in the oxidizable phase coinciding with organic and sulfur compounds. Arsenic is present in the phase 1 for about 10% of the total content, in station S_7, phase 1 percentage rises up to 16%. The lower land alluvial basin of the Ganges River is recognized as an arsenic-affected area. Arsenic in solution probably is easily entrapped in the fine grained organic-rich sediments deposited in the Ganges delta [101]. The percentage of silt (lower than 70% except in S_1 and S_2) may have contributed to a low retention of dissolved as since coarse sediments are less efficient at retaining as.

Cadmium was mainly present in the labile phase (more than 60%) in all the stations with the exception of station S_7, where the Cd labile percentage represents only 25% of the total concentration. Cadmium concentrations were negligible in phases 2 and 3. The highest labile Cd concentration was measured at station S_3, the closest to the city of Calcutta. Datta and Subramanian [96] found that the concentrations of elements in the non-detrital phases were higher in stations sampled in the Hugli River around Calcutta than in samples collected along Brahmaputra and Meghna rivers. The petroleum refinery, industrial, and mining effluents carried by the Hugli River may be responsible for this higher concentrations of non-detrital fractions.

About 40% of the total Cu concentration is associated to the residue, while 33% of Cu is bound to Fe-Mn oxide and hydroxide (phase 2). The high percentage of Cu in the residue is likely due to the fact that Cu is easily chemisorbed on or incorporated in clay minerals [103]. All the samples showed lower Cu concentrations in exchangeable phase, with percentage ranging from 7% (S_7) to 22% (S_5), with a mean of 15%. Copper is characterized by high complex constant with organic matter thus it can be hypothesized that Cu is bound to labile organic matter such as lipids, proteins, and carbohydrates. On the other hand, high-element concentration in labile phase could be related to recent coastal input [39].

Manganese was found in all the four sediment phases, as observed by other researchers [45, 104]. Manganese is the most mobile element since it is present with the highest percentage (a mean of 42%) in the labile phase. This is probably because of the known close association of Mn with carbonates [105] as endorsed by other workers [69, 106]. In this phase, weakly sorbed Mn retained on sediment surface by relatively weak electrostatic interactions may be released by ion exchange processes and dissociation of Mn-carbonate phase [2]. The result indicates that considerable amount of Mn may be released into environment if conditions become more acidic [107]. The highest Mn labile percentage was measured in S_6 (57%). Differently, in S_7, Mn in the residue represents 65% of the total concentration, while the labile Mn is only 15%. A substantial Mn percentage was also found in the residue (mean 37.8%), followed by the reducible phase (14.7%), in which Mn exists as oxides and may be released if the sediment is subjected to more reducing conditions [108].

The major geochemical phase for Pb in these sediments was the Fe-Mn oxides phase (mean 55.7%) followed by the residual phase (mean 30.2%) while lower percentage of the total Pb are bound to exchangeable-labile (mean 5.3%) and oxidizable phases (mean 6.8%). At S_3 (Babughat), the reducible part is as high as 65% and only 19.9% of the total is associated with the residue. Atmospheric input as fallout from vehicular emission can be probably the major input of Pb for this station. The relatively high percentage of Pb in reducible phase is in agreement with the known ability of amorphous Fe–Mn oxides to scavenge Pb from solution [109, 110]. Caille et al. [111] observed that resuspension of anoxic sediment results in a rapid desorption of Pb and Cu adsorbed to sulphides. Thus, a high element percentage in the reducible fraction is a hazard for the aquatic environment because Fe and Mn species can be reduced into the porewaters during early diagenesis by microbially mediated redox reactions [112]. Dissolution will also release Pb associated with oxide phases to the porewater possibly to the overlying water column [113] and to benthic biota [79]. The major sources of Pb are from intensive human activities, including agriculture in the drainage basin [114], auto exhaust emission together with atmospheric deposition [115]. In addition, a substantial contribution from the factories located in the upstream of the Hugli River dealing with Pb producing lead ingots and lead alloys play a vital role as referred by Sarkar et al. [116].

The percentage of Zn in residue is highly variable (38.5–70%) and the distribution pattern in each fraction showed the following order: residual>reducible>oxidizable> exchangeable and bound to carbonates. There was some difference in Zn speciation among the sampling sites: in stations S_1, S_2, and S_3 about 40% of Zn is present in the residue, while in the other stations this percentage increases to more than 60%. In station S_1, the exchangeable and oxidisable phases shared over 22% of the total Zn, whereas labile Zn was as low as 4.6% at S_7. A major part of Zn (16.3%) is associated with Fe–Mn oxide phase, because of the high stability constants of Zn oxides. Iron oxides adsorb considerable quantities of Zn and these oxides may also occlude Zn in the lattice structures [117].

The BCR procedure as discussed above showed satisfactory recoveries, detection limits, and standard deviations for determinations of heavy metals/metalloid in the sediments. It is evident from the present results of the fractionation studies that the metals/metalloids in the sediments are bound to different fractions with different strengths leading to variations in mobility and availability and some of them show significant spatial variations subject to diverse environmental stresses. This type of association between metals and the sediments can be understood in detail by sequential extraction techniques. Hence the application of sequential extraction is fully justified as the quantification of different forms of metal is more meaningful than the estimation of its total metal concentrations. The strength values can, therefore, give a clear indication of sediment reactivity, which in turn assess the risk connected with the presence of metals in this wetland environment. The results obtained suggest the need for corrective remediation measures due to the higher accumulation of potentially dangerous metals/metalloids, which in most cases exceed the limits established by certain legislation.

Comparison with Sediment Quality Guidelines

Results obtained after total and sequential extraction are compared with Sediment Quality Guidelines (SQGs). Table 6 reports consensus-based values, such as TEC (concentration below which harmful effects on sediment-dwelling organisms were not expected) and PEC (concentration above which harmful effects on sediment-dwelling organisms were expected to occur frequently), and effect range-low and range-medium, such as ERL (concentrations below which adverse biological effects were observed in less than 10% of studies) and ERM (concentrations above which effects were more frequently observed in more than 75% of studies).

Table 6: Sediment Quality Guidelines concentrations with respect to total and labile element concentrations found in the analyzed samples (expressed as µg/g of dry weight)

Element	Phase	Si<TEC	TEC	Si<TEC<PEC	PEC	Si<ERL	ERL	ERL<Si<ERM	ERM
As	Total	All	9.79	None	33	S5,S6,S7,S8	8.2	S1,S2,S3,S4,S9	70
	Labile	All		None		All		None	
Cd	Total	S1,S2,S4,S5,S6,S7,S8,S9	0.99	S3	4.98	S1,S2,S4,S5,S6,S7,S8,S9	1.2	S3	9.6
	Labile	S1,S2,S4,S5,S6,S7,S8,S9		S3		S1,S2,S4,S5,S6,S7,S8,S9		S3	
Cr	Total	None	43.4	All	111	All	81	None	370
	Labile	All		None		All		None	
Cu	Total	S1,S4,S5,S7,S8	31.6	S2,S3,S6,S9	149	S1,S3,S4,S5,S6,S7,S8	34	S2,S9	270
	Labile	All		None		All		None	
Ni	Total	None	22.7	All	48.6	None	20.9	All	51.6
	Labile	All		None		All		None	
Pb	Total	All	35.8	None	128	All	46.7	None	218
	Labile	All		None		All		None	
Zn	Total	All	121	None	459	All	150	None	410
	Labile	All		None		All		None	

Comparing our results with the SQGs, it is revealed that for Pb and Zn in all the stations the measured concentrations are lower than both TEC and ERL. As regards Cd, concentration measured in station S_3 is higher than TEC and ERL but lower than PEC and ERM both in term of total and labile concentration. For this station, some possible toxic effect on benthic organism can be hypothesized, in particular because of the large amount of element bound to the most labile phase of the sediment. Considering Cu, some stations (S_2, S_3, S_6, and S_9) exhibit total concentrations higher than TEC but lower than PEC. Concentrations of Cu are higher than ERL but lower than ERM only in stations S_2 and S_9. Since only 7–22% of total Cu is bound to the labile phase, in all stations Cu labile concentrations are lower than TEC and ERL. Total As concentrations in stations S_1, S_2, S_3, S_4, and S_9 are higher than ERL value but lower than TEC value since more than 50% of total. As is not found in the residue, attention should be paid to a change in the environment conditions which could induce a release of as from the sediments. Total Ni and Cr concentrations are higher than TEC (Ni is also higher than ERL) but lower than PEC (and ERM in the case of Ni) in all the stations. Nevertheless, more than 70% of Ni as well as 90% of Cr are present in the residual fractions, therefore adverse impacts on organisms is very much negligible.

Mean sediment quality guidelines quotients (mSQGQ) have been developed for assessing the potential effects of contaminant mixtures in sediments [118]: they are determined by calculating the arithmetic mean of the quotients derived by dividing the concentrations of chemicals in sediments by their respective SQGs. The probability of observing sediment toxicity can be estimated by comparing the mSQGQ in a sample to previously published probability tables. It is important to keep in mind that mSQGQs cannot be used to accurately predict the uptake and bioaccumulation of sediment-bound chemicals by fish, wildlife, and humans, even if there is considerable evidence that this assessment tool can be predictive of the presence or absence of toxic effects [118].

SQGQs are calculated for seven elements considering ERM as sediment quality guidelines (Table 7). The mean quotient values ranges from 0.16 in station S_5 to 0.24 in station S_3. Using PEC values instead of ERM, the mean SQGQ ranges from 0.25 in station S_5 to 0.38 in station S_3 (Table 7).

Table 7: Mean Sediment Quality Guidelines Quotients calculated for the nine stations using PEC and ERM as SQGs

Stations	$SQG_{QPE}C$	$SQG_{QER}M$
S1	0.30	0.19
S2	0.33	0.21
S3	0.38	0.24
S4	0.33	0.21
S5	0.25	0.16
S6	0.28	0.18
S7	0.32	0.21
S8	0.28	0.18
S9	0.34	0.22

Compiled data from multiple data sets reporting 10-day toxicity test conducted on amphipod species in saltwater showed that the incidence of toxicity for a range of SQGQ of 0.25–0.5 is ~35%, while for a mean SQGQ range from 0.1 to 0.25, the incidence of toxicity lowers to ~20%. Measures recorded in a survey of Biscayne Bay (port of Miami and the adjoining saltwater reaches of the lower Miami River, FL, USA) showed that the average amphipod survival (*Ampelisca abdita*) decreased slightly from the least contaminated (ERMQ <0.03) to the intermediate category, (ERMQ included in 0.03–0.2 range) then decreased greatly in the most contaminated sediments (ERMQ included in 0.2–2 range). Therefore, we can presume a low toxicity of sediments sampled in the nine stations for benthic organisms. It is important to note that the benthic response to contaminants covaried among stations with both the mean ERM quotients and the effect of natural factors, such as the sediment texture, TOC, and salinity [118].

Statistical Analyses

The relationships between variables and the differences between stations were evaluated by PCA. The analysis was performed on 36 objects (four sediment phases for nine stations) and 11 variables (Al, As, Cd, Co, Cr, Cu, Fe, Mn, Ni, Pb, Zn). Two significant components were identified explaining 68.3% and 14.5% of the total variance, respectively. By studying the loadings of the variables (Figure 2a) on the components it can be seen that all the elements except Cd, Mn, and Pb are significantly correlated.

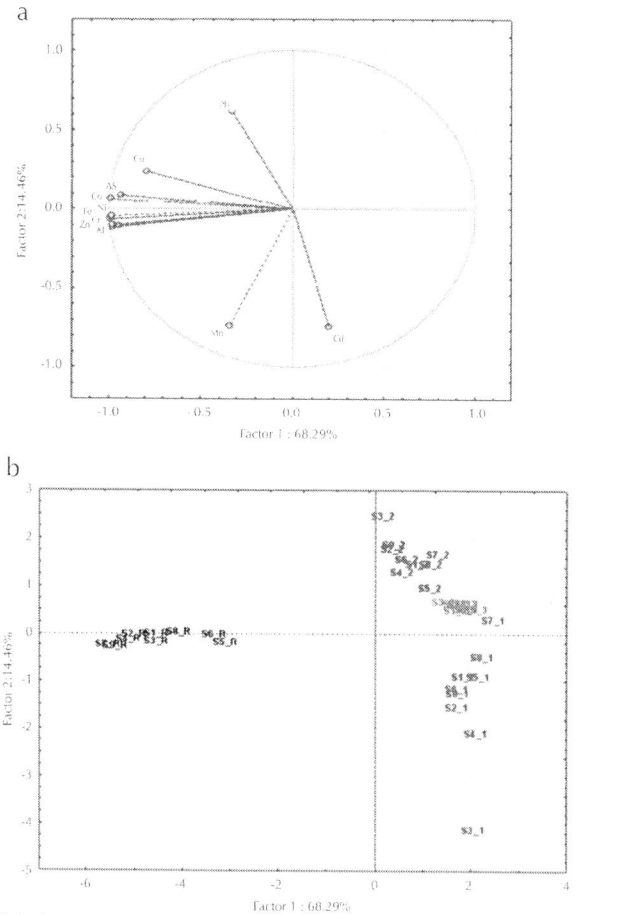

Figure 2: Principal component analysis: a) variable plot; b) score plot (phase 1, 2, 3 and 4, corresponding to labile, reducible, oxidizable and residue, are identifiable by 1, 2, 3 and 4 suffix and different colors in the score plot).

Unlike the other elements, most of Cd and Mn are present in the first phase: labile Cd and Mn represent more than 60% and 40% of the total concentration, respectively, except in station S_7. Cadmium and Mn speciation can be ascribed to their considerable affinity for carbonates.

Lead is the only element which is bound to the reducible phase for more than 50%. Lead is a very reactive element in water column and, having scavenging type behavior, is easily bound to hydroxy- and oxyligands. Copper is positively and significantly correlated with all elements except Cd and Mn, but with lower correlation coefficients (0.66–0.81).

In the score plot (Figure 2b) phases 1, 2, 3, and 4 (corresponding to labile, reducible, oxidizable, and residual phases respectively) are identifiable by 1, 2, 3, and 4 suffix, respectively. In all stations, residue concentrations were characterized by negative values of PC1 and consequently by high concentrations of Al, As, Co, Cr, Fe, Ni, and Zn. Conversely, in the positive PC1 semi-axis labile and oxidizable metal concentrations, which represent a small percentage of the total elements, are distributed. For all stations, reducible concentrations are distributed along the positive PC2 semi-axis, i.e., high Pb concentrations, with a maximum for station S_3 and a minimum for S_5. The group formed by elements bound to organic matter and sulphides (phase 3) is characterized by low values of both PC1 and PC2. Therefore, a low percentage of elements (higher than 20% exclusively for all as data and for Zn in stations S_1 and S_3) are bound to the oxidizable phase, suggesting the presence of an oxidant environment. High Mn and Cd concentrations are associated with negative values of PC2, therefore a relatively high concentration of labile Mn and Cd is present in all samples (in particular in S_3), except station S_7. Samples are prevalently grouped in relation to the sediment geochemical phase, suggesting a similar element speciation among the stations. Station S_7 represents an exception; in fact the labile fraction is closely associated to the oxidizable phase group.

A HCA was carried out by applying Euclidean distances to quantitatively identify specific groups of similar stations. In the dendrogram of the sampling stations (Figure 3), we can note two main clusters: the first represented by station S_7, characterized by the highest element percentage bound to residue, and the second constituted by all the remainder stations. In the second group, a subgroup formed by station S_5 and S_6 can be individuated.

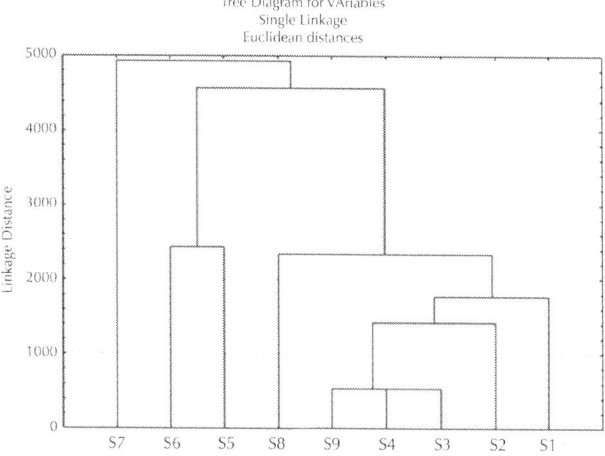

Figure 3: Dendrogram indicating linkage of sites on the basis of element concentrations.

Station S_7 was sampled in a marine coastal environment; it is characterized by a peculiar grain size percentage respect to the other stations, being the sand percentage as high as 98.6%. In general, the concentrations of elements are much higher in fine than in coarse fraction because the fine fraction larger specific surface facilitates absorption processes. As previously noted by Ramirez et al. [39], this pattern is particularly evident for Cd. It is interesting to note that the marine coastal stations S_8 and S_9 are more similar to river stations than to station S_7. Both the stations S_8 and S_9 are, in fact, located in front of the Hugli river runoff, while station S_7 is located easternmost and probably is less influenced by the Hugli river discharge.

Conclusion and Recommendation

The study provides valuable information on the potential mobility of trace elements in sediments collected along the stretch of Hugli River and in the Sundarban mangrove wetland (northeastern part of the Bay of Bengal). The results obtained adopting BCR sequential extraction method provided the following important information: (i) Al, Cr, and Fe were found mostly in the residual phase while the other elements were found in the four phases sharing different proportions; (ii) the dominant Cd, Mn, and Pb proportion was found in the non-residual fractions and

(iii) Mn had the highest percentage in the labile phase. This is worthwhile to mention that coastal environment of West Bengal is considerably constrained due to implementation of dredging, construction of port/harbor and other industrial activities. The authors strongly recommend for periodical monitoring on the bioavailability and mobility of trace elements, control the mixing of effluent of the concentration of heavy metals in the region, environmental remediation, treatment of industrial effluent and municipal wastewater for effective management of this estuarine system. It is wisely suggested that an environmental recovery framework should be urgently implemented to avoid extension of heavy metal contamination (especially As).

CASE STUDIES FROM OTHER COASTAL REGIONS IN INDIA

Although the importance of metal speciation and fractionation has been realized in developed countries, the subject has not really taken off in India and only few references are available on the speciation of metals in Indian rivers. Speciation of selected heavy metals geochemistry in surface sediments (n=10 was studied by Venkatramanan et. al. [119] from Tirumalairajan river estuary, east coast of India. The results obtained from sequential extraction showed that a larger portion of the metals were associated with the residual phase, although they are available in other fractions too.

Trace metal fractionation in the Pichavaram mangrove–estuarine sediments in southeast coast of India was studied by Ranjan et. al. [120] considering the pronounced changes due to occurrence of tsunami (2004) A 5-step sequential extraction procedure was applied to assess the effects of tsunami on mobility and redistribution of selected elements (Cd, Cr, Cu, Fe, Mn, Ni, Pb, and Zn) in coastal sediments revealed that metals in the residual fraction (lattice bound) had the highest concentration suggesting their non-availability and limited biological uptake in the system. Majority of the metals (except Mn) do not constitute a risk based on the different geochemical indices.

Fractionation of selected metals in the sediments of Cochin estuary and Periyar River (southwest coast of India) was studied by Mohan et. al. [121] The results reveal that remobilization potential of metals bound is in the range of low to medium risk to various sedimentary phases is different and is based on bond strength. Therefore, the strength values can give a clear indication of sediment reactivity that can be used to assess the risk related with metals to the aquatic organisms.

RISK ASSESSMENT CODE (RAC)

The risk assessment code (RAC) mainly applies the sum of the exchangeable and carbonate bound fractions for assessing the availability of metals in sediments. These fractions are considered to be weakly bonded metals which may equilibrate with the aqueous phase and thus become more rapidly bioavailable [11, 33]. This is important because the fractions introduced by anthropogenic activities, such as agricultural runoff and tourism, are typified by the adsorptive, exchangeable, and bound to carbonate fractions, which are weakly bonded metals that could equilibrate with the aqueous phase and thus become more rapidly bioavailable [122]. According to RAC guideline (Table 8), for any metal, soil/sediment which can release in exchangeable and carbonate fractions, less than 1% of the total metal will be considered safe for the environment and soil/sediment with 11-30% carbonate and exchangeable fractions will be at medium risk to the environment. On the contrary, soil/sediment releasing in the above fractions more than 50% of the total metal has considered being highly dangerous, which can be easily enter the food chain [123].

Table 8: Criteria of Risk Assessment Code [123]

Grade	Exchangeable and bounded to carbonate metal (%)	Risk
I	<1	No risk
II	1 – 10	Low risk
III	11 – 30	Medium risk
IV	31 – 50	High risk
V	>50	Very high risk

Heavy-Metal Fractionation in surface sediments was studied by Dhanakumar et. al. [124] in the Cauvery river estuarine region, southeastern coast of India The results revealed that most of the samples fall under the category from low- to high-risk class and from low-risk to very high-risk class in terms of labile fractions of Pb as well as Zn and Cu, respectively.

CONCLUSIONS

From the above discussion it is revealed that geochemical fractionation approach to the chemical speciation has provided a useful tool and opens a new dimension in assessing the potential mobility/ bioavailability of heavy metals and metalloids in soils/sediments and opens a new dimension in the field of ecology and environmental chemistry. More efficient, non-laborious and time saving processes techniques in this field of chemical speciation are also coming up to get valid information regarding geochemical behavior of soils/sediments. Besides geochemical fractionation, Dezileau et al. [125] opined that total Fe or Fe/Al may be used to infer millennial-scales climate changes in the south eastern pacific while performing sequential extraction of Fe in marine sediments from the Chileau continental margin. However, the chemical partitioning should be carefully used in the assessment of environmental pollution as large amount of metals may naturally occur as anthropogenic fractions (including loosely bonded ions, sulfide ions and metals associated with sediments).

ACKNOWLEDGEMENTS

The authors gratefully acknowledge full support and cooperation of the Springer press, UK for extending permission in publishing the research paper of the journal Environmental Monitoring & Assessment, vol. 184(12), pp:7561-77, 2012.

This study was partially supported by the European Fund for Economic and Regional Development (FEDER) through the Program Operational Factors of Competitiveness (COMPETE) and National Funds through the Portuguese Foundation for Science and Technology (PEST-C/MAR/UI 0284/2011, FCOMP 01 0124 FEDER 022689).

REFERENCES

1. Gleyzes C, Tellier S, Astruc M. Fractionation studies of trace elements in contaminated soils and sediments: a review of sequential extraction procedures. Trends in Analytical Chemistry 2002; 21 (6) 451-467.

2. Tessier A, Campbell PGC, Bisson M. Sequential extraction procedure for the speciation of particulate trace metals. Analytical Chemistry 1979; 51 844-851.

3. Sahuquillo A, Rigol A, Rauret G. Overview of the use of leaching/ extraction tests for risk assessment of trace metals in contaminated soils and sediments. Trends in Analytical Chemistry 2003; 22(3) 152-159.

4. Calmano W, Hong J, Förstner U. Binding and mobilization of heavy metals in contaminated sediments affected by pH and redox potential. Water Science and Technology 1993; 28(8/9) 53-58.

5. Eggleton J, Thomas KV. A review of factors affecting the release and bioavailability of contaminants during sediment disturbance events. Environment International 2004; 30 973-980.

6. Ure AM, Quevauviller P, Muntau H, Griepink B. Speciation of heavy metals in solids and harmonization of extraction techniques undertaken under the auspices of the BCR of the Commission of the European Communities. International Journal of Environmental Analytical Chemistry 1993; 51 135-151.

7. Florence TM. The speciation of trace elements in waters. Talanta 1982; 29 345-69.

8. Lung W. Speciation analysis – why and how? Fresenius' Journal of Analytical Chemistry 1990; 337 557-564.

9. Schroeder WH. Development in the speciation of mercury in natural waters. Trends in Analytical Chemistry 1989; 8 339-342.

10. Rauret G, Rubio R, Lopez-Sanchez JF, Cassassas E. Determination and speciation of copper and lead in sediments of a Mediterranean River (River Tenes, Catalonia, Spain). Water Research 1988; 22(4) 449-51.

11. Pardo R, Barrado E, Perez L, Vega M. Determination and association of heavy metals in sediments of the Pisuerga River. Water Research 1990; 24(3) 373-379.

12. Jardo CP, Nickless G. Chemical association of Zn, Cd, Pb and Cu in soils and sediments determined by the sequential extraction technique. Environmental science & Technology Letters 1989; 10 743-52.

13. Tessier A, Campbell PGC, Bisson M. Heavy metal speciation in the Yamaska and St. Francois rivers (Quebec). Canadian Journal of Earth Sciences 1980; 17 90-105.

14. Elsokkary LH, Muller G. Assessment and speciation of chromium, nickel, lead and chromium in the sediments of the river Nile, Egypt. Science of the Total Environment 1990; 97/98 455-63.

15. Ure AM. Single extraction schemes for soil analysis and related applications. Science of the Total Environment 1996; 178(1/3) 3-10.

16. Rauret G. Extraction procedures for the determination of heavy metals in contaminated soil and sediment. Talanta 1998; 46(3) 449-455.

17. Quevauviller PH, Lachica M, Barahona E, Rauret G, Ure A, Gomez A, Muntau H. Interlaboratory comparision of EDTA and DTPA procedures prior to certification of extractable trace elements in calcareous soil. Science of the Total Environment 1996; 178(1/3) 127-132.

18. Quevauviller PH, van der Slootr HA, Ure A, Muntau H, Gomez A, Rauret G. Conclusions of the workshop: harmonization of leaching/extraction tests for environmental risk assessment. Science of the Total Environment 1996; 178(1/3) 133-139.

19. Bacon JR, Davidson CM. Is there a future for Sequential chemical extraction? Analyst 2008; 133 25-46

20. Förstner U, Lechsber R, Davis RA, Hermitte PL. (eds.), Chemical methods for assessing bioavailable Metals in Sludges. Elsevier, London, 1985.

21. Meguellati M, Robbe D, Marchandise P, Astruc M, Proceedings International Conference on Heavy Metals in the Environment, Heidelberg CEP Consultants, Edinburgh, 1983, p. 1090.

22. Filgueiras AV, Lavilla I, Bendicho C. Chemextraction for metal partitioning in environmental solid samples, Journal of Environmental Monitoring 2002; 4 832-857.

23. Lopez-sanchez JF, Rubio R, Rauret G. Comparison of two sequential extraction procedures for trace metal partitioning in sediments. International Journal of Environmental Analytical Chemistry 1993; 51 113-121.

24. Hlavay J, Polyak K. Chemical speciation of elements in sediment samples collected at Lake Balaton. Microchemical Journal 1998; 58 281-290.

25. Tokalioglu S, Kartal S, Elçi L. Determination of heavy metals and their speciation in lake sediments by flame atomic absorption spectrometry after a four-stage sequential extraction procedure, Analytical Chimica Acta 2000; 413 33-40.

26. Belazi AU, Davidson CM, Keating GE, Littlejohn D. Determination and speciation of heavy metals in sediments from the Cumbrian coast, NW England, UK. Journal of Analytical Atomic Spectrometry 1995; 10 233-240.

27. Mester Z, Cremisini C, Ghiara E, Morabito R. Comparison of two sequential extraction procedures for metal fractionation in sediment samples. Analytical Chimica Acta 1998; 259 133-142.

28. Gomez-Ariza JL, Giraldez I, Sanchez-Rodas D, Morales E. Comparison of the feasibility of three extraction procedures for trace metal partitioning in sediments from southwest Spain, Science of Total Environment 2000; 246 271-283.

29. Zhang T, Shan X, Li F. Comparison of two sequential extraction procedures for speciation analysis of metals in soils and plant availability. Communications in Soil Science and Plant Analysis 1998; 29 1023-1034.

30. Albores AF, Cid BP, Gomez P, Lopez EF. Comparison between sequential extraction procedures and single extrations for metal partitioning in sewage sludge samples. Analyst 2000; 125 1353-1357.

31. Ho MD, Evans GJ. Operational speciation of cadmium, copper, lead and zinc in the NIST standard reference materials 2710 nad 2711 (Montana soil) by the BCR sequential extraction procedure and flame atomic spectrometry. Analytical Communications 1997; 34 353-364.

32. Zemberyova M, Bartekov J, Hagarov I. The utilization of modified BCR three-step sequential extration procedure for the fractionation of Cd, Cr, Cu, Ni, Pb and Zn in soil reference materials of different origins, Talanta 2006; 70 973-978.

33. Singh KP, Mohan D, Singh VK, Malik A. Studies on distribution and fractionation of heavy metals in Gomti river sediments- a tributary of the Ganges India. Journal of Hydrology 2005; 312 14-27.

34. Davidson CM, Duncan AL, Littlejohn D, Ure AM, Garden LM. A critical evaluation of the three-stage BCR sequential extraction procedure to assess the potential mobility and toxicity of heavy metals in industrially-contaminated land. Analytica Chimica Acta 1998; 363(1) 45-55.

35. McLaren RG, Crawford D. Studies on soil copper I. The fractionation of copper in soils. Journal of Soil Science 1973; 24(2), 172-181.

36. Kersten M, Förstner U. Speciation of trace elements in of water-soluble organic components. Sediments and combustion waste. In Ure AM, Davidson CM (ed.) Chemical speciation in the environment. Blackie Academic and Professional, Glasgow, UK. 1995; 234-275.

37. Ianni C, Magi E, Rivaro P, Ruggieri N. Trace metals in Adriatic coastal sediments: distribution and speciation pattern. Toxicological and Environmental Chemistry 2000; 78 73-92.

38. Ianni C, Ruggieri N, Rivaro P, Frache R. Evaluation and comparison of two selective extraction procedures for heavy metal speciation in sediments. Analytical Sciences 2001; 17 1273-1278.

39. Ramirez M, Massolo S, Frache R, Correa J. Metal speciation and environmental impact on sandy beaches due to El Salvador copper mine, Chile. Marine Pollution Bulletin 2005; 50 62-72.

40. Borovec Z, Tolar V, Mraz L. Distribution of some metals in sediments of the central part of the Labe (Elbe) River, Czech Republic. Ambio 1993; 22 200-205.

41. Campanella L, D'Orazio D, Petronio BM, Pietrantonio E. Proposal for a metal speciation study in sediments. Analytica Chimica Acta 1995; 309 387-393.

42. Zdenek B. Evaluation of the concentrations of trace elements in stream sediments by factor and cluster analysis and the sequential extraction procedure. Science of the Total Environment 1996; 177 237-250.

43. Gomez-Ariza JL, Giraldez I, Sanchez-Rodas D, Moralesm E. Metal sequential extraction procedure optimized for heavily polluted and iron Oxide rich sediments. Analytica Chimica Acta 2000; 414 151-164.

44. Lopez-Sanchez JF, Sahuquillo A, Fiedler HD, Rubio R, Rauret G, Muntau H, Quevauviller P. CRM 601, a stable material for its extractable content of heavy metals. Analyst 1998; 123 1675-1677.

45. Usero J, Gamero M, Morillo J, Gracia I. Comparative study of three sequential extraction procedures for metals in marine sediments. Environment International 1998; 24 478-496.

46. Martin R, Sanchez DM, Gutierrez AM. Sequential extraction of U, Th, Ce, La and some heavy metals in sediments from Ortigas River, Spain. Talanta 1998; 46 1115-1121.

47. Agnieszka S, Wieslaw Z. Application of sequential extraction and the ICPAES method for study of the partitioning of metals in fly ashes. Microchemical Journal 2002; 72 9-16.

48. Templeton DM, Ariese F, Cornels R. IUPAC guidelines for terms related to chemical Speciation and Fractionation of elements. Pure and Applied Chemistry 2001; 72 1453-1470.

49. Bordas F, Bourg ACM. A critical evaluation of sample for storage of Contaminated Sediments to be investigated for the potential mobility of their heavy metal load. Water. Air. Soil. Pollution 1998; 103 137-149.

50. Kersten M, Frostner U. Chemical fractionation of heavy metals in anoxic estuarine and coastal sediments. Water Science and Technology 1986; 18 121-130.

51. Dold B. Speciation of the most soluble phases in a sequential extraction procedure adapted for geochemical studies of copper sulfide mine waste. Journal of Geochemical Exploration 2003; 80 55-68.

52. Favas PJC, Pratas J, Gomes MEP, Cala V. Selective chemical extraction of heavy metals in tailings and soils contaminated

by mining activity: Environmental implications. Journal of Geochemical Exploration 2011; 111 160-171.

53. Pardo R, Vega M, Debán L, Cazurro C, Carretero C. Modelling of chemical fractionation patterns of metals in soils by two-way and three-way principal component analysis. Analytica Chimica Acta 2008; 606 26-36.

54. Cuong TD, Obbard JP. Metal speciation in coastal marine sediments from Singapore from Singapore using a modified BCR-sequential extraction procedure. Applied Geochemistry 2006; 21 1335-1346.

55. Dapaah RK, Takano N, Ayame A. Solvent extraction of Pb (II) from acid medium with zinc Hexamethylenedithiocarbamate followed by back-extraction and subsequent determination by FAAS. Analytica Chimica Acta 1999; 386 281-286.

56. Gomez-Ariza JL, Giraldez I, Sanchez-Rhodes D, Morales E. Metal readsorption and re-distribution during analytical fractionation of trace elements in toxic estuarine sediments. Analytica Chimica Acta 1999; 399 295-307.

57. Cheam V, Lechner J, Sekerka I, Desrosiers R, Nriagu J. Development of laser- excited atomic fluorescence spectrometer and a method for the direct determination of lead in Great Lake waters. Analytica Chimica Acta 1992; 269 129-136.

58. Fischer E, Van D, Berg CMG. Anodic Stripping Voltammetry of Pb and Cd using a Hg film electrode and thiocyanate. Analytica Chimica Acta 1999; 385 273-280.

59. Morales MM, Mart P, Llopis A, Compos L, Sagrado S. An environmental study by factor Analysis of surface sea waters in the Gulf of Valencia (Western Mediterranean). Analytica Chimica Acta 1999; 394 109-117.

60. Hirade M, Chen Z, Sugimoto K, Kawaguchi H. Co precipitation with tin (IV) hydroxide followed by removal of tin carrier for the Determination of trace heavy metals by graphite-furnace atomic absorption Spectrometry. Analytica Chimica Acta 1980; 302 103-107.

61. Ridout PS, Jones HR, Williams JG. Determination of trace elements in a marine reference material of lobster hepatopancreas (TORT-1) using inductively coupled plasma mass spectrometry. Analyst 1988; 113 1383-1386.

62. Sakao SY, OgawaY, Uchida H. Determination of trace elements in seaweed samples by inductively coupled plasma mass spectrometry. Analytica Chimica Acta 1999; 355 121-127.

63. Baeyens W, Monteny F, Leermakers M, Bouillon S. Evalution of sequential extractions on dry and wet sediments. Analytical and Bioanalytical Chemistry 2003; 376 890- 901.

64. Guevara-Riba A, Sahuquillo A, Rubio R, Rauret G. Assessment of metal mobility in dredged harbour sediments from Barcelona, Spain. Science of the Total Environment 2004; 321 241-255.

65. Idris AM, Eltayeb MAH, Potgieter-Vermaak SS, Grieken R, Potgieter JH. Assessment of heavy metal pollution in Sudanese harbours along the Red Sea coast. Microchemical. Journal 2007; 87 104-112.

66. Wepener V, Vermeulen LA. A note on the concentrations and bioavailability of selected metals in sediments of Richards Bay Harbour, South Africa. Water SA 2005; 31 589-595.

67. Esslemont G. Heavy metals in seawater, marine sediments and corals from the Townsville section, Great Barrier Reef Marine Park, Queensland. Marine Chemistry 2000; 71 215- 231.

68. Coetzee PP. Determination and speciation of heavy metals in sediments of the Hartebeespoort Dam By sequential extraction. Water SA 1993; 19 291-300.

69. Salmons W, Förstner U. Trace metal analysis on polluted sediments. Part II:evaluation of Environmental impact. Environmental science and Technology letters 1980; 1 14-24.

70. Li X, Thornton I. Chemical partitioning of trace and major elements in soils contaminated by mining and smelting activities. Applied geochemistry 2000; 16 1693-1706.

71. Kiratli N, Ergin M. Partitioning of heavy metals in surface Black Sea sediments. Applied Geochemistry 1996; 11 775-788.

72. Rauret G, Lopez-Sanchez JF. New sediment and soil CRMs for extractable Trace metal content. International Journal of Environmental Analytical Chemistry 2001; 79 81-95.

73. Salmons W. Adoption of common schemes for single and sequential extractions of Trace metals in soil and sediments. International Journal of Environmental Analytical Chemistry 1993; 51 3-4.

74.   Fiedler HD, Lopez-Sanchez JF, Rubio R, Rauret G, Quevauviller PH. Study of the stability of extractable trace metal contents in a river sediment using Sequential extraction. Analyst 1994; 119 1109-1114.

75.   Ramos L, Hernandez LM, Gonzalez MJ. Sequential fraction of copper, lead, copper, Cadmium and zinc in soils from or near Donana National Park. Journal of Environmental Quality 1994; 23 7-50.

76.   Tu Q, Shan XZ, Ni Z. Evaluation of a sequential extraction procedure for the Fractionationation of amorphous iron and manganese oxides and organic matter in soils. The Science of The total Environment 1994; 151 159-165.

77.   Wang S, Jia Y, Wang S, Wang X, Wang H. Fractionation of heavy metals in shallow marine sediments from Jinzhou Bay, China. Journal of Environmental Science (China) 2010 22 23-31.

78.   Yuan CG, Shi JB, He B, Liu JF, Liang LN. Speciation of heavy metals in marine sediments from the East China Sea by ICP-MS with sequential extraction. Environment International 2004; 30 769-783.

79.   Jones B, Turki A. Distribution and Speciation of heavy metals in surficial sediments from the Tees Estuary, North – East England. Marine Pollution Bulletin 1997; 34 768-779.

80.   Ajay SO, Van Loon GW. Studies on redistribution during the analytical fractionation of metals in sediments. The Science of the Total Environment 1989; 87 171-187.

81.   Pempkowiak J, Sikora A, Biernacka E. Speciation of heavy metals in marine sediments vs their bioaccumulation by mussels. Chemosphere 1999; 39 313- 321.

82.   Forstner U, Wittmann GTW. Metal Pollution in the Aquatic Environment, Springer- Verlag, Berlin. Springer-Verlag, Heidelberg; 1981.

83.   Fedotov PS, Zavarzina, a AG, Spivakov BYa, Wennrich, b R, Mattusch J, De K, Titzeb PC, Demin VV. Accelerated fractionation of heavy metals in contaminated soils and sediments using rotating coiled columns, Journal of Environmental Monitoring 2002; 4 318-324.

84. Nemati K, Kartini N, Bakar A, Abas MR, Sobhanzadeh E. Speciation of heavy metals by modified BCR sequential extraction procedure in different depths of sediments from Sungai Buloh, Selangor, Malaysia, Journal of Hazardous Materials 2011; 192(1) 402-410.

85. Mossop KF, Davidson CM. Comparison of original and modified BCR sequential extraction procedures for the fractionation of copper, iron, lead, manganese and zinc in soils and sediments. Analytica Chimica Acta 2003; 478 (1) 111-118.

86. Walkey A, Black TA. An examination of the Dugtijaraff method for determining soil organic matter and proposed modification of the chronic and titration method. Soil Science 1934; 37 23-38.

87. Farnham IM, Johannesson KH, Singh AK, Hodge VF, Stetzenbach KJ. Factor analytical approaches for evaluating groundwater trace element chemistry data. Analytica Chimica Acta 2003; 490 123-138.

88. Ragno G, De Luca M, Ioele G. An application of cluster analysis and multivariate cassification methods to spring water monitoring data. Microchemical Journal 2007; 87 119-127.

89. Jonathan MP, Ram Mohan V. Heavy metals in sediments of the inner shelf off the Gulf of Mannar, Southeast coast of India. Marine Pollution Bulletin 2003; 46 263-268.

90. Sunil Kumar R. Distribution of organic carbon in the sediments of Cochin mangroves, south west coast of India. Indian Journal of Marine Science 1996; 25 274-276.

91. Janaki-Raman D, Jonathan MP, Srinivasalu S, Armstrong-Altrin J S, Mohan SP, Ram-Mohan V. Trace metal enrichments in core sediments in Muthupet mangroves, SE coast of India: Application of acid leachable technique. Environmental Pollution 2007; 145 245-257.

92. Sarkar SK, Bilinski SF, Bhattacharya A, Saha M, Bilinski H. Levels of elements in the surficial estuarine sediments of the Hugli river, northeast India and their environmental implications. Environment International 2004; 30 1089-1098.

93. Tam NFY, Wrong YS. Spatial variation of heavy metalsin surfe sediments of Hong, Kong mangrove swamps. Environmental Pollution 2002; 110 195-205.

94. Subramanian V, Mohanachandran G. Heavy metals distribution and enrichment in the sediments of southern east coast of India. Marine Pollution Bulletin 1990; 21 324-330.

95. Chatterjee M, Massolo S, Sarkar SK, Bhattacharya AK, Bhattacharya BD, Satpathy KK, Saha S. An assessment of trace element contamination in intertidal sediment cores of Sunderban mangrove wetland, India for evaluating sediment quality guidelines. Environmental Monitoring and Assessment 2009; 150 307-322.

96. Datta DK, Subramanian V. Distribution and fractionation of heavy metals in the surface sediments of the Ganges–Brahmaputra–Meghna river system in the Bengal Basin. Environmental Geology 1998; 36 93-101.

97. Müller G. Schwermetalle in den sedimenten des Rheins-Veranderungen seit. Umschan Verlag 1979; 79 133-149.

98. Salomon W, Förstner U. Metals in the hydrocycle. Berlin: Springer; 1984.

99. Buccolieri A, Buccolieri G, Cardellicchio N, Dell'atti A, Leo AD, Maci A. Heavy metals in marine sediments of Taranto Gulf (Ionian Sea, Southern Italy). Marine Chemistry 2006; 99 227-235.

100. Dowling CB, Poreda RJ, Basu AR, Aggarwal PK. Geochemical study of arsenic release mechanisms in the Bengal Basin groundwater. Water Resources Research 2002; 38(9) 1173-1190.

101. Acharyya SK, Lahiri S, Raymahashay BC, Bhowmilk A. Arsenic toxicity of groundwater in parts of the Bengal basin in India and Bangladesh: the role of Quaternary stratigraphy and Holocene sea-level fluctuation. Environmental Geology 2000; 39 231-238.

102. Takarina ND, Browne DR, Risk MJ. Speciation of heavy metals in coastal sediments of Semarang, Indonesia. Marine Pollution Bulletin 2004; 49 854-874.

103. Pickering WF. Metal ion speciation—soil and sediments (a review). Ore Geology Reviews 1986; 1 83-146.

104. Ngiam LS, Lim PE. Speciation patterns of heavy metals in tropical estuarine anoxic and oxidized sediments by different sequential extraction schemes. Science of the Total Environment, 2001; 275 53-61.

105. Dassenakis M, Adrianos H, Depiazi G, Konstantas A, Karabela M, Sakellari A, Scoullos M. The use of various methods for the study of metal pollution in marine sediments, the case of Euvoikos Gulf, Greece. Applied Geochemistry 2003; 18 781-794.

106. Morillo J, Usero J, Gracia I. Heavy metal distri-bution in marine sediments from the southwest coast of Spain. Chemosphere 2004; 55 431-442.

107. Thomas RP, Ure IS Davidson CM, Littlejoh D, Rauret G, Rubio R, López-Sánchez JF. Three-stage sequential extraction procedure for the deter-mination of metals in river sediments. Analytica Chimica Acta 1994; 286 423-429.

108. Panda D, Subramanian V, Panigrahy RC. Geochemical fractionation of heavy metals in Chilka Lake (east coast of India) – a tropical coastal lagoon. Environmental Geology 1995; 26 199-210.

109. Dawson EJ, Macklin MG. Speciation of heavy metals in floodplain and flood sediments: a reconnaissance survey of the Aire Valley, West Yorkshire, Great Britain Environmental Geochemistry and Health 1998; 20 67-76.

110. Ramos L, González M, Hernández L. Sequential extraction of copper, lead, cadmium, and zinc in sediments from Ebro River (Spain): relationship with levels detected in earthworms. Bulletin of Environmental Contamination and Toxicology 1999; 62 301-308.

111. Caille N, Tiffreau C, Leyval C, Morel JL. Solubility of metals in anoxic sediment during prolonged aeration. Science of the Total Environment 2003; 301 239-250.

112. Canfield DE. Reactive iron in marine sediments. Geochimica et Cosmochimica Acta 1989; 53 619-632.

113. Petersen W, Wallman K, Li PL, Schroeder F, Knauth HD. Exchange of trace elements at the sediment– water interface during early diagenesis processes. Marine and Freshwater Research 1995; 46 19-26.

114. Monbet P. Mass balance of lead through a small macrotidal estuary: the Morlaix River estuary (Brittany, France). Marine Chemistry 2006; 98 59-80.

115. Adriano DC. Trace elements in terrestrial environments. New York: Springer; 1986.

116. Sarkar SK, Saha M, Takada H, Bhattacharya A, Mishra P, Bhattacharya B. Water quality management in the lower stretch of the river Ganges, east coast of India: an approach through environmental education. Journal of Cleaner Production 2007; 15 1559-1567.

117. Banerjee ADK. Heavy metal levels and solid phase speciation in street dusts of Delhi, India. Environmental Pollution 2003; 123(1) 95-105.

118. Long ER, Ingersoll CG, MacDonald DD. Calculation and uses of mean sediment quality guideline quotients: a critical review. Environmental Science and Technology 2006; 40 1726-1736.

119. Venkatramanan S, Ramkumar T, Anithamary I, Jonathan MP. Speciation of selected heavy metals geochemistry in surface sediments from Tirumalairajan river estuary, east coast of India. Environmental Monitoring and Assessment 2013; 185(8) 6563-6578.

120. Ranjan RK, Singh G, Routh J, Ramanathan AL. Trace metal fractionation in the Pichavaram mangrove–estuarine sediments in southeast India after the tsunami of 2004. Environmental Monitoring and Assessment 2013 (article in press).

121. Mohan M, Augustine T, Jayasooryan KK, Chandran MSS, Ramasamy EV. Fractionation of selected metals in the sediments of Cochin estuary and Periyar River, southwest coast of India, Environmentalist 2012; 32 383-393.

122. Hseu ZY. Extractability and bioavailability of zinc over time in three tropical soils incubated with biosolids. Chemosphere 2006; 63 762-771.

123. Perin G, Craboledda L, Lucchese M, Cirillo R, Dotta L, Zanette ML, Orio AA. Heavy metal speciation in the sediments of Northern Adriatic Sea- a new approach for environmental toxicity determination, in: T.D. Lekkas (Ed.), Heavy Metal in the Environment 1985; 2 454-456.

124. Dhanakumar S, Murthy KR, Solaraj G, Mohanraj R. Heavy-Metal Fractionation in Surface Sediments of the Cauvery River Estuarine Region, Southeastern Coast of India, Arch Environmental Contamination Toxicology 2013; 65 14-23.

125. Dezileau L, Pizarro C Rubio MA. Sequential extraction f iron in marine sediments from the Chilen continental margin. Marine Geology 2007; 241 111-116.

6

Heavy Metal Pollution in Soil and Water in Some Selected Towns in Dunkwa-on-Offin District in the Central Region of Ghana as a Result of Small Scale Gold Mining

Jerome D. A. Kpan[1,2], Boadu Kwesi Opoku[2], and Anukwah Gloria[2]

[1]College of Optical-Electrical and Computer Engineering, University of Shanghai for Science and Technology, Shanghai, China
[2]Department of Chemistry, University of Cape Coast, Cape Coast, Ghana

ABSTRACT

Illegal small scale gold mining popularly called "Galamsey" in our local communities is on the in- crease. This has led to concerns about the level of environmental pollution resulting from these mining activities. This work was conducted to determine the level of heavy metal contamination in the environment due to the activities of the small scale miners. This paper discusses the concen- trations of some selected heavy metals—Hg, Pb, and Cu which were measured in 14 sampling sites in Dunkwa-on-Offin in the Central Region of Ghana, known for these activities. The heavy metal concentrations have been investigated for soil and water samples in the selected towns and com- pared with the relevant guidelines of the Environmental Protection Agency. The concentration of heavy metals was measured by using AAS. In most locations, the concentration for the investigated heavy metals far exceeded the concentration admitted by the guidelines. The mean concentration of Lead was 95.13 mg/Kg for soil and 190.27 mg/L in water; Copper was 63.26 mg/Kg in soil and 75.92 mg/L in water while Mercury was 140.87 ug/Kg in soil and 211.31 mg/L in water. The mean recorded concentrations in the sensitive areas exceeded greatly. Hence the levels of heavy metal contamination have spread beyond control.

INTRODUCTION

Small scale gold mining, popularly called "galamsy" in Ghana and other developing countries, is seen as a source of subsistence and a determinant of the environmental degradation [1]. In Ghana, small scale gold mining is set to be responsible for about 5% of the annual gold production. However, this gold mining of late has become un- popular as it is seen as the source of significant heavy metal contamination of the environment [2]. Some of the impacts associated with this small scale gold mining include the destruction of vegetation, land degradation and the pollution of water bodies. The rate at which mercury is discharged into the environment and water bodies is very alarming [3] - [5]. In small scale gold mining, simple tools are used in the recovery of gold from the land. Pits are dug haphazardly and these remain uncovered long after their operations. It's reported that in a prospecting work in a field, a pit that was dug revealed the presence of mercury in the soil [6]. Tetteh, et al. (2010) [7] reported high levels of mercury and zinc content in the top soil of towns in Wassa West. The levels of the concentration, however, decreased with distance from the main mining centers and extended beyond most probably due to aerial dispersion of the metal from mining areas.

Amalgamation is the preferred method used for the extraction of this free gold from its ores. The gold amalgam is usually roasted to release the mercury and to concentrate the gold. The excess mercury which is discarded into the environment finds its way into the water bodies [8]. Methylmercury in water and mercury oxides in the air as a result of the gold amalgam finds its way in humans through the food chain as in ingestion of mercury contaminated food or fish and through inhalation [2] [9] .

The main goal of this paper is to determine the levels/concentration of trace metals introduced into the envi- ronment as a result of the small scale mining activities and to compare the values with that of the Romanian guideline which are being used by the Ghana Environmental Protection Agency as a standard to determine if it has reached pollution levels. It will also provide evidences whether the mining activities actually introduced the trace metals into the environment. Lastly, this will help to determine situation and suggest appropriate environmental strategies to contain the problem coupled with recommendations to control the problem.

MATERIALS AND METHODS

Study Area and Location of Sampling Sites

The study was conducted in Dunkwa-on-Offin in the Dunkwa District in the Central Region of Ghana. The Dis- trict shares boarders with the Eastern and Ashanti Regions of Ghana. The samples investigated were soil and water collected from various parts of the study area. Fourteen (14) sampling sites were located (Tables 1-4). Sites 1 to 12 were located in farms within the vicinity of Dunkwa-on-Offin in various directions and distances from the centre of the town while site 13 was located in Nkronya, the main site of the small scale mining activities. Site 14 was located in the botanical gardens of the University of Cape Coast.

Table 1: Name and shortest distance of the sites 1 to 3 from the centre of the town (In South Direction of Nkronya)

Name of Village	Distance (Km)
Akyempem	3.00
Aduman	10.00
Kyekyewere	12.50

Table 2: Name and shortest distance of the sites 4 to 6 from the centre of the town (In North-Western Drection of Nkronya)

Name of Village	Distance (Km)
Atekyem	1.50
Kwameprakrom	6.40
Ayemfori	12.75

Table 3: Name and shortest distance of the sites 7 to 9 from the centre of the town (In South-Western Direction of Nkronya)

Name of Village	Distance (Km)
Mfuom	2.50
Babiaraneha	4.25
Asikuma	9.25

Table 4: Name and shortest distance of the sites 10 to 12 from the centre of the town (In North-Western Direction of Nkronya)

Name of Village	Distance (Km)
Acquakrom	1.00
Tikyakrom	2.25
Manukrom	5.50

Ghana Environmental Protection Agency has adopted Romanian guideline on levels of trace metals in water and in soil as a standard to determine if the determined levels have reached pollution stage [10].

Sample Collection and Analysis

The water and soil samples were randomly collected from 14 sampling sites in 14 different areas within Dunkwa and the University of Cape Coast in the Central Region of Ghana. The sampling sites were located in various directions and distances from the centre of the town.

At each soil sample collection site, the samples were collected to cover the plough zone. The samples were collected by removing the top litter first and with a Teflon-coated soil auger; the sample was collected into an already well washed plastic containers and sealed. They were then conveyed to the laboratory for analysis.

For the water, samples were collected at each site from the river which runs through all the towns selected in the Dunkwa District. The sample bottles were rinsed with deionised water twice before samples were collected. Collected samples were then preserved with 0.5 ml of concentrated nitric acid and stored in an ice chest with a temperature of 4°C.

The samples were then analysed for the various trace metals. Atomic absorption spectrophotometer was em- ployed in the analysis of the selected metals, Hg, Pb and Cu. The cold vapour technique was used for Hg deter- mination while in the determination of Pb and Cu, the flame-AAS technique was used [11] [12] .

RESULTS AND DISCUSSIONS

Although there are clear guidelines of limit values for maximum metal concentrations in drinking water, air and food, there is still no equivalent consensus of permissible levels of metals in rivers, sediments and soils by the Ghana Environmental Protection Agency. This is as a result of uncertainties of metal does-relationships in soil and sediments and hence some contradictory values.

The results were compared with similar studies conducted in Romani with similar characteristics. The WHO guideline values for drinking water and that of US for soil and sediments were also used for comparison [13] .

The total metal concentrations recorded in this study clearly indicate that the situation of metal pollution within the study area where this small scale gold mining is taking place has reached intervention levels. The concentra- tions of Mercury in particular were identified and discussed because of its immediate implications on human health.

Heavy Metal Pollution in Soil

The 3 metals were detected in the soil sample at varying concentrations in all the selected towns. Table 5 shows the mean concentrations of the trace metals that were found in the soil sample. The pollutant concentration of Cu and Pb were measured in mg/Kg while that of Hg was in µg/Kg.

Table 5: Distribution of heavy metal concentrations in the soil sample

Town	Distance from Nkronya (Km)	mg Pb/ Kg		mg Cu/ Kg		µg Hg/ Kg	
		Mean	Std. D	Mean	Std. D	Mean	Std. D
Kyekyewere	3.00	99.80	9.30	51.60	0.10	261.50	13.05
Aduman	10.00	63.40	0.70	64.30	0.10	137.50	5.15
Akyempem	12.50	86.20	0.60	45.70	0.15	248.20	11.90
Atekyem	1.50	30.40	0.25	85.40	0.20	148.20	0.60
Kwameprakrom	6.40	89.40	8.85	49.90	0.10	133.70	9.85
Ayamfori	12.75	132.00	0.45	71.40	0.15	146.90	5.15
Mfuom	2.50	58.10	10.15	72.80	0.05	83.30	5.85
Babiaraneha	4.25	82.40	3.10	59.90	0.20	74.00	13.85
Asikuma	9.50	76.46	2.20	66.10	0.55	69.80	1.25
Acquahkrom	1.00	25.00	2.35	66.10	0.44	116.90	10.00
Tikyakrom	2.25	86.20	2.70	52.10	0.07	110.50	9.35
Manukrom	5.50	46.10	3.65	38.10	0.06	136.20	6.15
Nkronya	0.00	154.00	0.25	90.40	0.05	265.70	3.80
UCC	275.00	6.20	0.00	3.10	0.00	1.10	0.00
Standard		80.00		50.00		40.00	

From Figure 1, the lead concentrations in the various towns depicted a varying trend. About 50% of the towns showed values less than the standard. The towns which showed the higher value concentration were towns where there has been extensive cultivation of cocoa and other similar food crops. Here, there has been profilic use of various pesticides, herbicides and artificial fertizers to boost crop yeilds. This could cause the increase of the metal in the soil.

The concentration of copper as seen in Figure 2 demonstrated the higher concentration of the metal. This could have come about as a reuslt of the area ever housing a copper mine dating back to 1493 - 1600 [14].

The total mercury concentrations were higher in all the towns as shown in Figure 3. There has been increased illegal mining activities across the entire district and with corresponding increase in mercury in the environment. This really suggests that the small scale gold mining and recovery activities contribute to the level of mercury. Comparing this results to that obtained on the sample from the botanical garden of the University of Cape Coast and that of an organized gold mine in Obuasi [15], the illegal mining indeed caused the pollution of the environment with the trace metals.

Heavy Metal Pollution in Water Sample

The heavy metals were detected in the water sample at varying concentrations in all the towns. Table 6 shows the mean concentrations of the trace metals in the water sample. The pollutant concentration of Cu and Pb were meas- ured in mg/L while that of Hg was in µg/L.

From Figures 4-6, the mean trace metal concentrations depicted a varying trend from the different towns. However, one thing has been observed that the concentrations were well above the standard level. The highest levels were recorded in towns that are located upstream of the river Offin which run through all the towns under study. Most of the concentrations reduced with stretch of the water flow down the stream. The river Offin which runs through Dunkwa district is up stream of its source in Tarkwa in the Western Region of Ghana which is well known for its gold mining activities as far back as 1471 [14] . The trace metal concentration however, confirmed the idea that heavy metal pollutants and others in water turn to reduce along the distance of travel for the moving surface of a water body [8] [16] [17] .

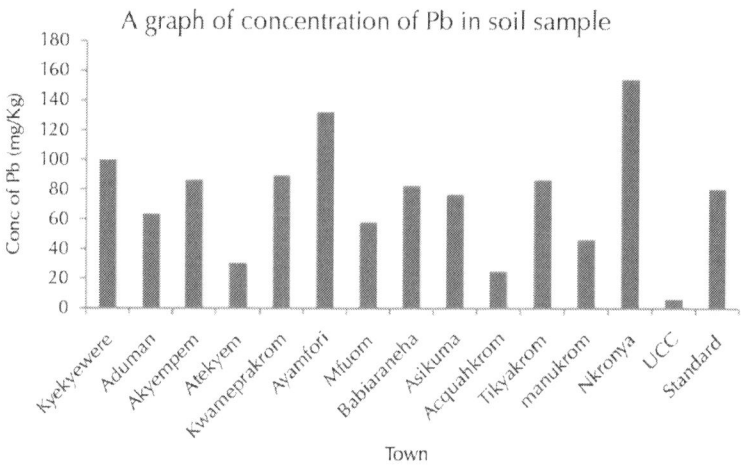

Figure 1: A graph showing the distribution of Pb concentrations in the soil sample at the various towns.

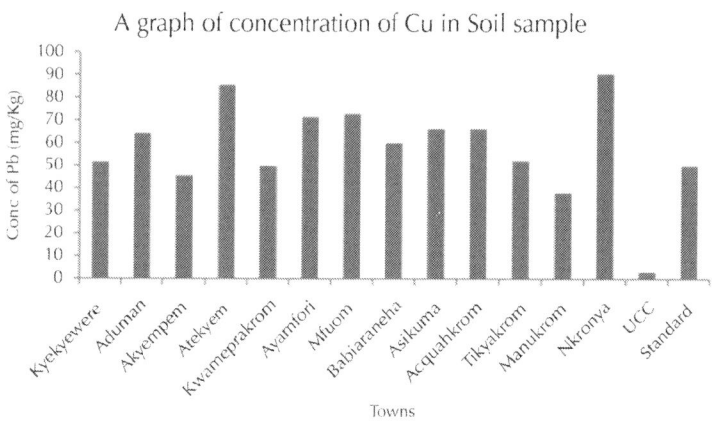

Figure 2: A graph showing the distribution of Cu concentrations in the soil sample at the various towns.

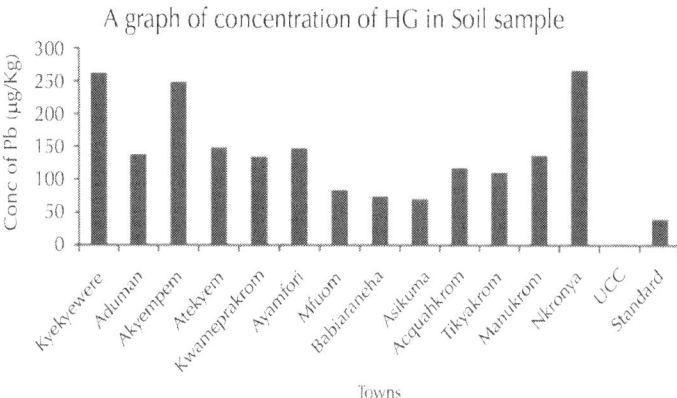

Figure 3: A graph showing the distribution of Hg concentrations in the soil sample at the various towns.

The results obtained from the sample taken from the botanical garden of the University of Cape Coast were expected because there is no mining activity taking place in the garden or even within the environs of the Uni- versity. The presence of some trace metals as seen could be due largely to background levels in the soil and water as a result of the use of fertilizers and also carried by wind from elsewhere.

STRATEGIES AND RECOMMENDATION

Strategies for Environmental Control

The government, the civil society and nongovernmental organizations in Ghana have formed a joint public initi- ative to control the environmental pollution caused by small scale mining. These strategies should entail a joint initiative of the small scale miners, the government, civil societies, non-governmental organizations, and finally environmental entities and experts.

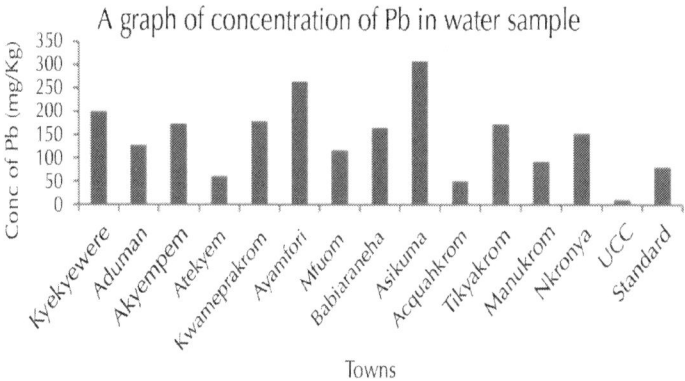

Figure 4. A graph showing the distribution of Pb concentrations in the water sample at the various towns.

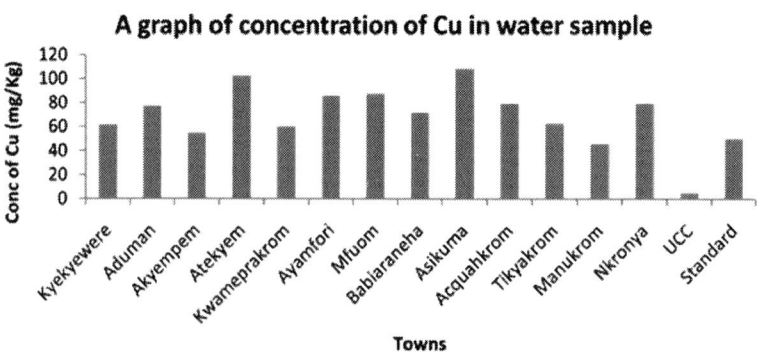

Figure 5. A graph showing the distribution of Cu concentrations in the water sample at the various towns.

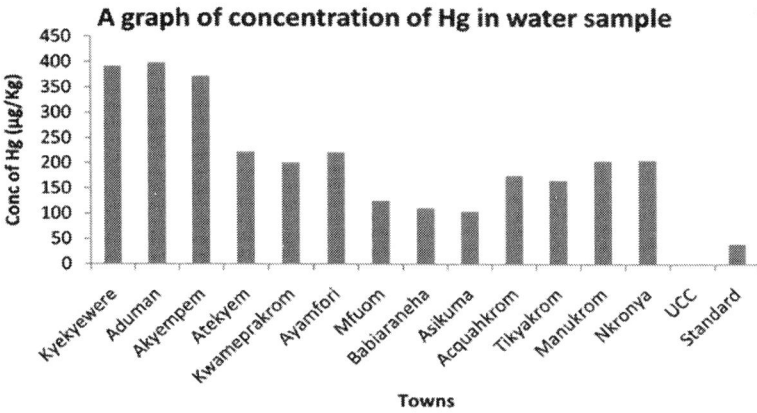

Figure 6: A graph showing the distribution of Hg concentrations in the water sample at the various towns.

Revising environmental laws to ensure that they are relevant to the situation at hand, the laws should recog- nize and prescribe the best mining practices that promote environmental conservation.

Implementation of strong rules and regulations that are aimed at controlling the activities of the small scale gold miners; the regulations should provide for punishment and penalties for those who are found culpable.

Small scale miners should be encouraged to avoid sluicing of gold in the open air, rivers beds and streams. The small scale miners should avoid burning the gold amalgam in the open air as this releases dangerous fumes of mercury and carbon dioxide.

The government should implement land reclamation program. The initiative should entail planting of more trees on the areas that have undergone deforestation. Planting of more trees ensures that the soil regains its fer- tility and also reduces soil erosion.

Table 6: Distribution of heavy metal concentrations in the water sample

Town	Distance from Nkronya (Km)		mg Pb/L		mg Cu/L		µg Hg/L	
		Mean	Std. D	Mean	Std. D	Mean	Std. D	
Kyekyewere	3.00	199.60	9.30	61.60	0.10	392.25	10.44
Aduman	10.00	126.80	0.70	77.16	0.10	398.55	4.12
Akyempem	12.50	172.40	0.60	54.84	0.15	372.30	9.52
Atekyem	1.50	60.80	0.25	102.48	0.20	222.30	0.48
Kwameprakrom	6.40	178.80	8.85	59.88	0.10	200.55	7.88
Ayamfori	12.75	264.00	0.45	85.68	0.15	220.35	4.12
Mfuom	2.50	116.20	10.15	87.36	0.05	124.95	4.68
Babiaraneha	4.25	164.80	3.10	71.88	0.20	111.00	11.08
Asikuma	9.50	308.00	2.20	108.48	0.55	104.70	1.00
Acquahkrom	1.00	50.00	2.35	79.32	0.44	175.35	8.00
Tikyakrom	2.25	172.40	2.70	62.52	0.07	165.75	7.48
Manukrom	5.50	92.20	3.65	45.72	0.06	204.30	4.92
Nkronya	0.00	152.92	0.25	79.32	0.05	206.25	3.04
UCC	275.00	10.00	0.01	5.00	0.00	0.02	0.00
Standard		80.00		50.00		40.00	

The miners should be encouraged to avoid land excavation by using modern techniques and tools in mining.

Recommendations

To control environmental pollution caused by the activities of the small scale gold miners, this research there- fore proposes the following recommendations.

The government should empower the Minerals commission to exercise its mandate through adequate funding and support.

The commission should enact legislations to control the small scale mining in the country.

The Minerals commission should conduct a proper and comprehensive research to determine the level of en- vironmental degradation caused by the activities of the small scale miners.

The players in the industry both small scale and large scale operators should adopt proper environmental management tools.

The commission should control illegal mining activities practiced by the small scale miners. The government should enable the licensed small scale operators in the mining sector access funding from financial institutions in the economy. Funding of their operations will enable them to adopt modern mining techniques and abandon traditional techniques such as shallow alluvial mining which leads to the massive deforestation and excavation of the earth surface. Funding will also enable the miners to procure modern mining tools which can enable them crush hard rocks containing gold without causing noise and dust.

The government, the Minerals Commission, the civil society, and non-governmental organizations should conduct proper campaigns to enhance awareness on the impacts of environmental pollution. The initiative should aim at educating the small scale miners on the impacts of their activities.

CONCLUSIONS

The study revealed that small scale mining activities indeed introduce mercury and other trace metal pollutants into the environment. The concentrations of these trace metals determined in this study showed that it has reached pollution levels in the environment and therefore need urgent attention. This is because the results as compared to the standards and other areas where organized gold mining has taken place indicated much higher concentrations of the selected trace metals.

REFERENCES

1.   Agyemang, I. (2010) Population Dynamics and Health Hazards of Small Scale Mining Activity in the Bolgatanga and Talensi-Nabdam Districts of the Upper East Region of Ghana. Indian Journal of Science and Technology, 3.

2.   Cobbina, S.J., Myilla, M. and Michael, K. (2013) Small Scale Gold Mining and Heavy Metal Pollution: Assessment of Drinking Water Sources in Datuku in the Talensi-Nabdam District. International Journal of Scientific & Technology Research, 2, 96.

3.   Manu, A., Twumasi, Y.A. and Coleman, T.L. (2004) Application of Remote Sensing and GIS Technologies to Assess the Impact of Surface Mining at Tarkwa, Ghana. Geoscience and Remote Sensing Symposium, IGARSS'04 Proceedings. IEEE Intrnational, 1, 572-574.

4.   Obiri, S. (2007) Determination of Heavy Metals in Boreholes in Dumasi in the Wassa West. District of Western Re- gion of the Republic of Ghana. Environmental Monitoring and Assessment, 130, 455-463.

5.   Tom-Dery, D., Dagben, Z.J. and Cobbina, S.J. (2012) Effect of Illegal Small Scale Mining Operations on Vegaetation Cover of Arid Northern Ghana. Research Journal of Environmental and Earth Sciences, 4, 674-679.

6.   Appiah, H. (1998) Organization of Small Scale Gold Mining Activities in Ghana. Journal of the South African Institute of Mining and Metallurgy, November 1998.

7.   Tetteh, S., Golow, A.A., Essumang, D.K. and Zugle, R. (2010) Levels of Mercury, Cadmium and Zinc in the Topsoil of Some Selected Towns in the Wassa West District of the Western Region of Ghana. Soil and Sediment Con- tamination: An International Journal, 19, 635-643.

8.   Ezeh (2007) Environmental Significance of Heavy Metals Distribution. Ebonyi River Drainage System, Abakaliki and Ohaozara Areas, South Eastern Nigeria. Ph.D. Thesis, Nnamdi Azikiwe University, Awka.

9.   Nartey, V.K., Klate, R.K., Hayford, E.K., Doamekpor, L.K. and Appoh, R.K. (2011) Assessment of Mercury Pollution in Rivers and Streams around Artisanal Gold Mining Areas of the Birim North District of Ghana. Journal of Environ- mental Protection, 2, 1227-1239.

10.   Olade (2009) Dispersion of Cd, Pb and Zn in Soils and Sediments of a Humid Tropical Ecosystem in Nigeria. In: Hut- chinson, T.C., Ed., Lead, Mercury, Cadmium and Arsenic in the Environment, John Wiley and Sons, New York.

11. Office of Research and Development, US Environmental Protection Agency (1991) Methods for the Determination of Metals in Environmental Samples. Office of Research and Development, US Environmental Protection Agency, Cin- cinnati, 293.

12. Perkin-Elmer (1993) Analytical Methods for Atomic Absorption Spectrophotometry. Norwalk, Connecticut.

13. WHO (2006) Guidelines for Drinking Water Quality. 4th Edition, World Health Organization, Geneva.

14. Hug, M.M. (1989) The Economy of Ghana: The First 25 Years Since Independence. Macmillian Press Ltd., London, 153-155.

15. Akabzaa, T.M., Banoeng-Yakubo, B.K. and Seyire, J.S. (2005) Impact of Mining Activities on Water in the Vicinity of the Obuasi Mine. 79, 377-379.

16. Ezeh, H.N. and Chukwu, E.A. (2011) Small Scale Mining and Heavy Metals Pollution of Agricultural Soils: The Case of Ishiagu Mining District, South Eastern Nigeria. Journal of Geology and Mining Research, 3, 87-104.

17. Kpan, J.D.A. (2008) Studies on Levels of Mercury, Zinc and Cadmium in Soils in the Vicinity of an Alluvial Goldmine at Dunkwa-on-Offin. Unpublished Thesis, University of Cape Coast, Cape Coast.

Culturable Heavy Metal-Resistant and Plant Growth Promoting Bacteria in V-Ti Magnetite Mine Tailing Soil from Panzhihua, China

Xiumei Yu[1], Yanmei Li[1], Chu Zhang[1], Huiying Liu[1], Jin Liu[1], Wenwen Zheng[1], Xia Kang[1], Xuejun Leng[1,2], Ke Zhao[1], Yunfu Gu[1], Xiaoping Zhang[1], Quanju Xiang[1], and Qiang Chen[1]

[1]Department of Microbiology, College of Resource and Environmental Sciences, Sichuan Agricultural University, Chengdu, China

[2]Environmental Monitoring Station, Panzhihua Municipal Environmental Protection Bureau, Panzhihua, China

ABSTRACT

To provide a basis for using indigenous bacteria for bioremediation of heavy metal contaminated soil, the heavy metal resistance and plant growth-promoting activity of 136 isolates from V-Ti magnetite mine tailing soil were systematically analyzed. Among the 13 identified bacterial genera, the most abundant genus was Bacillus (79 isolates) out of which 32 represented B. subtilis and 14 B. pumilus, followed by Rhizobium sp. (29 isolates) and Ochrobactrum intermedium (13 isolates). Altogether 93 isolates tolerated the highest concentration (1000 mg kg^{-1}) of at least one of the six tested heavy metals. Five strains were tolerant against all the tested heavy metals, 71 strains tolerated 1,000 mg kg^{-1} cadmium whereas only one strain tolerated 1,000 mg kg^{-1} cobalt. Altogether 67% of the bacteria produced indoleacetic acid (IAA), a plant growth-promoting phytohormone. The concentration of IAA produced by 53 isolates was higher than 20 µg ml^{-1}. In total 21% of the bacteria produced siderophore (5.50–167.67 µg ml^{-1}) with two Bacillus sp. producing more than 100 µg ml^{-1}. Eighteen isolates produced both IAA and siderophore. The results suggested that the indigenous bacteria in the soil have beneficial characteristics for remediating the contaminated mine tailing soil.

INTRODUCTION

Mining industry has caused extensive environmental and public health problems [1]–[3]. A wide variety of heavy metals such as zinc, lead, copper and cadmium have been detected in soil at mining sites presenting a major threat to the environment and population [4], [5]. Heavy metals cannot be biologically degraded and indefinitely persist in the environment. Heavy metals transferred through the food chain are a serious hazard to human health [6]. Due to contamination by heavy metals, mining sites are surrounded by large barren areas. The awareness of the detrimental heavy metal contamination at mining sites has increased in recent years.

The toxic heavy metals accumulated in soil can effectively impact the microbial community composition. Bacteria play an important role in maintaining soil fertility and structure. Because bacteria respond quickly and are sensitive to subtle environmental changes, they have been considered as efficient bio-indicators of soil quality [7]. Both the structural and functional bacterial diversity are important indicators of soil health [8]. Phytoremediation has been effectively used to remediate heavy metal-polluted sites as a sustainable remediation approach [9]. Plant-microbe partnerships may be utilized to improve biomass production and remediation [10]. Plant growth-promoting rhizobacteria (PGPR) that solubilize phosphate and synthesize growth-promoting substances such as indoleacetic acid (IAA) and siderophores can be applied in the plant-assisted bioremediation of metal-contaminated soil [11]–[14]. Phytoremediation utilizes heavy metal-tolerant plant species with metal accumulation ability. Since the addition of IAA to soil can enhance the uptake of metals in plant roots [15], [16], bacteria-producing IAA have been used to assist the phytoremediation of soil contaminated with heavy metals [17].

Metals such as iron, zinc, copper, manganese and nickel play important roles as essential or beneficial micronutrients of microorganisms [18], [19]. However, a high concentration of metal ions in soil shows serious effects on microbial communities by changing the community structure and decreasing diversity and total microbial biomass [13]. Therefore, microbial communities are useful indicators of the effect of contamination on soil health [20]. To have a functional role in remediation, bacteria in heavy metal-contaminated soil must first overcome the heavy metal stress. Microorganisms tolerate heavy metals by immobilizing metals on cell surfaces or transforming metals into less toxic forms, for example by precipitation, acidification and oxidation-reduction [21].

Panzhihua is an industrial and mining city in Sichuan of Southwest China with over 10^9 tonnes of ore reserves deposited as iron-vanadium-titanium oxide (V-Ti magnetite) [22]. The world class magmatic deposits of V-Ti magnetite in Panzhihua provide 20% of iron (Fe), 64% of vanadium (V) and 53% of titanium (Ti) supply for China [23]. Long-term mining activities have contaminated soils and sediments in Panzhihua by metals, especially by V, Ti and Fe [24]. The redox-sensitive vanadium is toxic to soil microorganisms and plants [25]. Even though titanium is beneficial to plants at low concentrations, high concentrations of titanium are toxic [26]. Iron is an essential nutrient serving as a catalyst for many cellular reactions, in particular those involving redox and O_2 chemistry [27]. More than 220 million m^3 of mine tailing has been piled up in Zhujiabaobao, Panzhihua, creating a serious environmental hazard. Because the mine tailing soil contains heavy metals, only few plants grow on it, creating large barren areas. Long term exposure to contaminant allows different bacteria to become adapted to the contaminant, making autochthonous bacteria more useful in bioremediating the contaminated environment compared to allochthonous ones [28]. Therefore, this study focuses on the culturable heavy metal-resistant and plant growth-promoting autochthonous bacteria from V-Ti magnetite mine tailing soil, with the aim of providing information for bioremediating the large area covered by the V-Ti magnetite mine tailing dam at Panzhihua.

MATERIALS AND METHODS

Study Site and Soil Sampling

Panzhihua (N26°05′~27°21′, E101°08′~102°15′) in Sichuan, China, is an important industrial and mining base with abundant mineral resources. Panzhihua V-Ti magnetite mine is by the Jinsha River in the southern part of Panxi rift valley upstream of the Yangtze River. The mining area includes six large scale iron deposits, numerous medium-sized coal, clay, dolomite and limestone deposits, and minor graphite, manganese and barite deposits [23]. Our study site did not involve endangered or protected species and provide a specific location, so no specific permissions were required for the location/activity.

Zhujiabaobao (N 26°37′2″, E 101°45′56″) in Panzhihua is a huge deposit of V-Ti magnetite. In the Zhujiabaobao mining area, there is more than 220 million m^3 of mine tailing and a large tailing dam. The ground around the tailing dam is barren and desert-like. Four soil samples were collected in May 2013 to a depth of 0–20 cm from the tailing dam by a five-point sampling method. Soil samples, fully mixed in sterile polyethylene bags, were kept at room temperature in the dark.

Chemical Analyses of Soil Samples

The soil samples were dispersed and passed through a 2 mm sieve before measuring pH, organic matter content and the concentrations of available nitrogen (N), phosphorus (P) and potassium (K) as described [29]. Soil available N was determined by Aalkali N-proliferation method, whereas and available P and K quantified with the ASI method. Soil organic matter was determined by the K_2CrO_7–H_2SO_4 oxidation method. To extract heavy metals, air-dried soil samples were passed through a 2 mm nylon sieve, and digested by 1:2:2 (V:V:V) HNO_3:HCl:$HClO_4$. Vanadium (V), titanium (Ti), iron (Fe), nickel (Ni), lead (Pb), zinc (Zn), manganese (Mn), copper (Cu), arsenic (As), cadmium (Cd) and chromium (Cr) were measured by inductively coupled plasma atomic emission spectroscopy (ICP-AES, IRIS Intrepid II, Thermo Electron corporation, USA) as described [30].

Isolation of Bacteria

A sample of 5 grams of soil was suspended in 45 ml sterile water with glass beads. After shaking for 30 minutes and letting settle for 5 minutes, 200 μl of the liquid phase was inoculated on beef extract-peptone agar medium (beef extract 3.0 g l^{-1}, peptone 10.0 g l^{-1}, NaCl 5.0 g l^{-1}, agar 18.0 g l^{-1}, pH 7.0) in Petri dish (90 mm diameter×10 mm depth). Isolates were selected based on differences in colony morphology and re-streaked several times on beef extract-peptone agar at 28°C until 136 pure cultures were obtained.

Genetic Identification of Isolated Bacteria

Total DNA was extracted by the phenol-chloroform method as described [31]. To group the 136 isolates, BOX-PCR with the primer BOXA1R (5'-CTACGGCAAGGCGACGCTGACG-3') was carried out as described [32], [33]. A dendrogram based on the BOXA1R-PCR fingerprints was drawn using Numerical Taxonomy and Multivariate Analysis System NTSYSpc 2.2 (Exeter Software, USA). An isolate from each of the 91 BOXA1R-PCR groups was chosen for 16S rRNA gene sequencing. Almost full length 16S rRNA gene was amplified by polymerase chain reaction (PCR) with the universal primers of 27F (5'-AGAGTTTGATCCTGGCTCAG-3') and 1492R (5'-GGTTACCTTGTTACGACTT-3') [34], [35] and sequenced at the Beijing Genomics Institute (Shenzhen, China). The sequences were submitted GenBank to assign accession numbers. The closest matching sequences were searched from GenBank with BLAST [36]. A neighbor joining the 16S rRNA phylogenetic tree was constructed using the neighbor joining method in MEGA 6.0 [37].

Heavy Metal Tolerance Tests

The resistance of the 136 isolates to lead, cadmium, zinc, copper, cobalt and nickel was assayed by spot-inoculating 10 μl of 10^8 cells ml^{-1} bacterial suspension on beef extract-peptone agar medium with the respective metal salts. $Pb(NO_3)_2$, $CdCl_2$, $ZnCl_2$, $CuSO_4$, $CoCl_2$ and $NiCl_2$ were added to the medium to obtain 200, 400, 600, 800 and 1000 mg kg^{-1} heavy metal concentrations. After incubation at 28°C for 5 days, the minimum inhibitory concentration (MIC) was defined as the lowest concentration of metal salt inhibiting bacterial growth. On the positive control plates without heavy metal the colonies were approximately four mm in diameter.

Indoleacetic Acid and Siderophore Production Assays

The plant growth-promoting activity of the isolates was evaluated by assaying the production of indole-3-acetic acid (IAA) and siderophores as described. In the qualitative IAA assay [38], isolates were grown in a beef extract-peptone liquid medium with 0.5 g l^{-1} tryptophan at 28°C and 140 rpm for 36 hours, 50 μl of the culture suspension was absorbed into a white porcelain board and, after adding 100 μl of the color reagent (4.5 g l^{-1} $FeCl_3$, 57.6% H_2SO_4), the board was incubated at 25°C for 30 min. A pink color indicated positive IAA production. A non-inoculated beef extract-peptone liquid medium with tryptophan served as a negative control. To quantify IAA production [38], [39], 4 ml of the color reagent was added to 2 ml of the culture supernatant obtained by centrifugation (8000 rpm for 5 min). Optical density at 550 nm was measured by spectrophotometry (WFJ2100, UNICO, China) after coloration for 30 min. IAA concentration in the supernatant was interpolated using an IAA standard curve. The distribution of IAA producers and heavy metal tolerant strains among the different taxa were compared with a Chi-square test.

In the qualitative siderophore assay, isolates were grown on chrome azurol sulphonate (CAS) agar to select siderophore producing strains [40]. To quantify siderophore production, siderophore producing strains were grown in Fiss minimal medium (5.03 g l^{-1} L-asparagine, 5.03 g l^{-1} KH_2PO_4, 5.0 g l^{-1} glucose, 0.5 mg l^{-1} $ZnCl_2$, 40 mg l^{-1} $MgSO_4$

and 0.5 μM FeSO$_4$) for two days. After centrifugation at 1000 g for 15 min, supernatant was mixed with CAS solution (1 vol: 1 vol) and incubated for 60 min. Optical density at 400 nm was measured by spectrophotometry (WFJ2100, UNICO, China) [41]. Siderophore concentration in the supernatant was interpolated using a deferoxamine mesylate salt (SIGMA, USA) standard curve. The IAA and siderophore production and heavy metal tolerance assays were done in triplicate.

RESULTS

Basic Physicochemical Properties of Soil Samples

To assess the quality of the V-Ti magnetite mine tailing soil, we first measured the soil physicochemical characteristics and heavy metal content in the soil (Table 1). The soil pH was low (5.28±0.91), as was the content of organic matter (16.98±4.45‰). Of the 103.50±36.84 mg kg^{-1} total N, approximately 13% was plant-available N. As expected, the iron, titanium and vanadium concentrations were high, up to 76.15, 28.19 and 5.58 g kg^{-1}, respectively. The manganese concentration was 1.42 g kg^{-1}. The concentration of chromium, zinc, copper and nickel were 98.51, 87.72, 56.75 and 48.11 mg kg^{-1}. In addition, lead (3.87 mg kg^{-1}), arsenic (0.94 mg kg^{-1}) and Cadmium (0.52±0.25 mg kg^{-1}) were detected in the mine tailing soil.

Table 1: The basic physicochemical properties and heavy metal concentrations in V-Ti magnetite mine tailing soil

Properties	Average value	Minimum value	Maximum Value VvalueVaaaValueVavalue
pH	5.28±0.91	4.48	6.34
Organic matter (‰)	16.98±4.45	12.72	22.29
Total N (mg kg^{-1})	103.50±36.84	72.85	153.73

Available N (mg kg⁻¹)	13.64±8.03	8.46	25.60
Available K (mg kg⁻¹)	11.71 ±3.86	7.21	15.98
Available P (mg kg⁻¹)	11.08±1.82	8.87	13.15
As (mg kg⁻¹)	0.94±0.58	0.47	1.77
Fe (mg kg⁻¹)	76145.84±3715.20	70782.50	78647.50
Ti (mg kg⁻¹)	28185.84±3264.46	23976.67	31596.67
V (mg kg⁻¹)	5584.38±2457.28	2974.00	7595.00
Cr (mg kg⁻¹)	98.51 ±9.20	90.61	111.72
Mn (mg kg⁻¹)	1417.17±141.02	1268.33	1543.75
Zn (mg kg⁻¹)	87.72±20.27	69.21	106.59
Cu (mg kg⁻¹)	56.75±30.65	30.05	95.58
Ni (mg kg⁻¹)	48.11±12.23	38.40	65.34
Pb (mg kg⁻¹)	6.90±1.37	5.62	8.84
Cd (mg kg⁻¹)	0.52±0.25	0.19	0.78

doi:10.1371/journal.pone.0106618.t001

Isolation of Bacteria and Genetic Identification

Based on differences in colony morphology, 136 bacterial strains were isolated from the V-Ti magnetite mine tailing soil. According to the BOX A1R-PCR fingerprint analysis, the similarities between the 136 isolates ranged from 0.54 to 1.00. Altogether there were 91 distinct fingerprint patterns. The 136 isolates were divided into two major groups, group I (81 isolates) and group II (55 isolates), at 54% similarity level One strain was chosen from each of the 91 distinct fingerprint pattern groups for 16S rRNA sequencing The sequences were assigned GenBank accession numbers KJ733935–KJ734025 The16S rRNA gene sequences of the 91 representative strains indicated that the group I and group II in the BOX A1R-PCR dendrogram represented Gram-positive and Gram-negative bacteria, respectively Seventy-nine isolates belonged to the genus Bacillus and represented eleven species Altogether 32 isolates were considered as representing B. subtilisand 14 as B. pumilus In addition to the Bacillus spp. isolates, isolates KT19 and KT84 were identified as Gram-positive strains displaying 99% similarity to the type strains of Paenibacillus tundrae and Microbacterium aerolatum, respectively

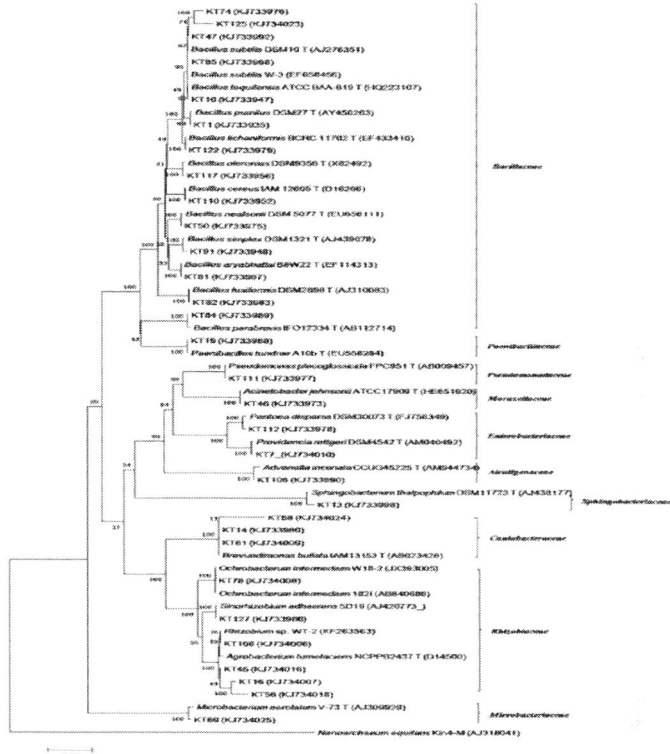

Figure 1: Neighbor-joining tree based on 16S rRNA sequences of isolated representative bacteria strains from V-Ti magnetite mine tailing soil by MEGA 6. The scale bar corresponds to 0.05 substitutions per nucleotide position. The numbers at nodes indicated the levels of bootstrap support (%) based on 1,000 resampled data sets; only values above 50% are given. Superscript "T" means type stains; Number in parentheses represents the sequence number in GenBank. Nanoarchaeum equitansKin4-M was used as an outgroup.

The 55 Gram-negative isolates were assigned to seven families and ten genera Altogether 43 of them belonged to the Rhizobiaceae and were assigned asRhizobium sp. (29 isolates), Ochrobactrum intermedium (13 isolates) and Sinorhizobium adhaerens (1 isolate).

Heavy Metal Tolerance of Isolated Bacteria

The heavy metal tolerance of the 136 bacterial isolates from the V-Ti magnetite mine tailing soil was determined as the minimum inhibitory concentration (MIC) (Figure 2). Most isolates showed MIC lower than 200 mg kg^{-1} for Zn (68.4%), Co (87.5%) and Ni (74.3%). Altogether 93 isolates tolerated the highest concentration (1,000 mg kg^{-1}) of at least one tested heavy metal; 71 strains tolerated 1,000 mg kg^{-1} cadmium whereas only one strain, Bacillus sp. KT-76, tolerated 1,000 mg kg^{-1} cobalt. Only three strains, the Rhizobium sp. KT27 and KT62, and theBacillus sp. KT43 displayed MIC less than 200 mg kg^{-1} for all the six heavy metals tested. Five strains, the B. licheniformis KT-87 and KT-88 and the Bacillus sp. KT-72, KT-74 and KT-76 were tolerant against all the tested heavy metals. B. licheniformis KT87 showed 1000 mg kg^{-1}MIC for Cd, Zn, Cu and Ni, 800 mg kg^{-1} MIC for Pb and 600 mg kg^{-1} MIC for Co. The MIC ofBacillus sp. KT72 for the six heavy metals was 1000 mg kg^{-1} (Pb, Cd, Ni) and 600 mg kg^{-1}(Zn, Cu, Co). When comparing the percentage of strains tolerant to four or more heavy metals, it was noted that, among the Rhizobium sp., multiple tolerant strains (34.9%) were less abundant than among the Bacillus spp. (54.4%) ($p<0.05$).

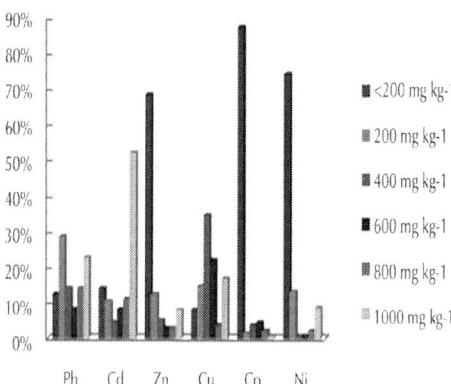

Figure 2: The minimum inhibitory concentrations (MICs) of bacterial isolates against six heavy metals. Pb, lead; Cd, cadmium; Zn, zinc; Cu, copper; Co, cobalt; Ni, nickel.

Indoleacetic Acid and Siderophore Production

Most of the isolates, altogether 91 strains, produced IAA ranging from 2.2 to 83.05 µg ml^{-1}. Eleven strains produced more than 60 µg ml^{-1} IAA. Six of these were Ochrobactrumsp., three were Bacillus spp., one was Providencia sp. and one Rhizobium sp. Ochrobactrumsp. KT80 produced highest concentration (83.05 µg mL^{-1}) of IAA among the 136 isolates, followed by Providencia sp. KT7 (79.02 µg mL^{-1}) and Rhizobium sp. KT24 (75.82 µg ml^{-1}).

Only 29 of the 136 isolates produced siderophore ranging from 5.50 to 167.67 µg ml^{-1}. All the six Rhizobium sp. isolates that produced siderophore were IAA producers, whereas only half of the eighteen Bacillus spp. and two Ochrobactrum sp. siderophore producers produced IAA. Bacillus sp. isolates KT93 and KT113 that did not produce IAA produced highest siderophore concentrations of 167.67 µg ml^{-1} and 105.33 µg ml^{-1}, respectively.

DISCUSSION

Basic Physicochemical Properties of the V-Ti Magnetite Mine Tailing Soil

The basic physicochemical properties are important factors for evaluating soil quality. The V-Ti magnetite tailing, a weakly acid soil, showed pH similar to that in an old Spanish Pb-Zn mine soil [42]. Soil pH is the best predictor of microbial diversity and community composition [43],[44]. The bacteria in the V-Ti magnetite mine tailing soil may prefer acid environment. The organic matter content of V-Ti magnetite mine tailing soil was lower than in agricultural and urban ecosystem soils [29], [45], but similar with other mine tailing area [42]. The V-Ti magnetite mine tailing also showed lower content of available N, P and K than agricultural and urban ecosystem soils [29], [45], The low contents of available N, P, K and organic matter suggested that V-Ti magnetite mine tailing soil was unfertile. The iron, titanium and vanadium concentrations were up to three, ten and 70 times higher than in US soils in average [46], respectively. The concentrations of

Cr, Zn, Cu, Ni and Mn were approximately 1.5 to 2.5 times higher than in US soils in average [46]. The concentrations of Fe, Cu and Zn were above values considered very high by Abreu et al [47] and the chromium concentration was almost twice as high as needed to inhibit alfalfa germination [48], plausibly explaining the scarce vegetation at the Zhujiabaobao V-Ti magnetite tailing dam. Therefore, phytoremediation of the barren V-Ti magnetite mine tailing soil should include increasing the content of available N, P, K and organic matter and lowing metal concentration.

Genetic Identification of Bacterial Isolates

To assess if the heavy metal-resistance and plant growth promoter-producing bacteria in the Zhujiabaobao V-Ti magnetite tailing dam soil would support phytoremediation, we isolated 136 bacterial strains, grouped them by BOX A1R-PCR and identified representative strains by 16S rRNA gene sequencing. The bacteria in the V-Ti magnetite mine tailing soil represented both Gram-negative and Gram-positive species. Bacillus spp. were the most abundant species, followed by Rhizobium spp. and Ochrobactrum spp. The spore and cyst forming capability ofBacillus spp. may explain why Bacillus spp. were abundant in the unfavorable environment of the V-Ti magnetite mine tailing soil. Autochthonous Bacillus from mine tailing in South Korea showed the ability to biomineralize heavy metals, such as Pb and Cr [49], [50]. Ochrobactrumspp. and Pseudomonas spp. have been used for the bioremediation of environmental pollutants[51]–[53]. These observations suggested that the indigenous bacteria might be useful for phytoremediation of the Zhujiabaobao mine tailing soil. The presence of multiple autochthonousRhizobium spp. implied that, with compatible leguminous plants, the rhizobium-legume symbiosis could be used to gradually increase nitrogen content and overall fertility in the barren soil. The symbiosis of rhizobia and leguminous plants has been effectively used to remediate contaminated soil [54], [55].

Heavy Metal Tolerance of Bacterial Isolates

To estimate the usefulness of the isolated bacteria in bioremediation, we assessed their heavy metal tolerance. Obviously, the bacteria in V-Ti magnetite mine soil have to tolerate the harsh environment polluted by heavy metals. The tolerance mechanisms include exclusion, extrusion, accommodation, bio-transformation and methylation or demethylation [13]. Bacteria can enhance metal solubility by producing acid and detoxify metals by removal, sequestering or immobilizing [56]. Since heavy metal tolerance is one of the most important factors for using an indigenous microorganism in bioremediation, recently the functional diversity of bacterial communities in contaminated soil has attracted more attention [7].

The bacterial isolates from V-Ti magnetite mine tailing soil showed diverse tolerance to different heavy metals. Most of the isolates did not tolerate the lowest tested concentration (200 mg kg^{-1}) of Ni, Co and Zn. The concentrations of Ni and Zn were low in the mine tailing soil. A few of the V-Ti magnetite mine tailing isolates, e.g. Bacillus sp. KT87 and Bacillus sp. KT72, showed tolerance to higher metal concentrations than isolates from a copper mine tailing, from a mercuric salt-contaminated soil and from chickpea rhizosphere soil [57]–[59]. Interestingly, even though the concentration of Cd in the mine soil was not high, more than half the isolates tolerated a high concentration of Cd (1,000 mg kg^{-1}). Likewise, even though the concentration of Pb in the mine tailing soil was low, some isolates tolerated a high concentration of Pb, i.e. the tolerance to heavy metals and the heavy metal content of soil did not directly correlate. Many mine tailing sites are polluted by multiple metals. For bacteria, the ability to survive, including variation in strains and characteristics, is related to environmental conditions and length of exposure to those conditions [60]. Therefore, the isolates from the V-Ti magnetite mine tailing soil showing multi-metal resistance were affected by the unfavorable environment. The culturable bacteria included isolates with multiple heavy metal tolerance, especially among theBacillus spp., suggesting that the indigenous bacteria are capable of assisting the bioremediation of the V-Ti magnetite mine tailing soil polluted by heavy metals.

Plant Growth-Promoting Activity of Bacteria

Plant-associated bacteria play a key role in host adaptation to changing environment by altering plant cell metabolism or promoting plant growth. Plant growth promoting rhizobacteria (PGPR) producing IAA and siderophore have been widely used to accelerate phytoremediation of metalliferous soil [13], [21]. Production of indoleacetic acid (IAA), a phytohormone, is a key characteristic of PGPR [61]. The addition of IAA to soil can enhance the uptake of metals in plant roots [13], [15], [62]. Even though PGPR are widely studied, few studies have systematically analyzed PGPR in contaminated soil. About 23% and 50% of Zn- and Cd-accumulating isolates from a former zinc and lead mining and processing site in Austria produced IAA and siderophore, respectively [63]. In V-Ti magnetite mine tailing soil the percentages of culturable rhizosphere IAA and siderophore producers were entirely different at 67% and 21%, respectively. The plant growth promoting activity of the isolates from the V-Ti magnetite mine tailing was stronger than that reported for Pseudomonas putida GR12-2 (IAA: 2.01 $\mu g\ ml^{-1}$) from the rhizosphere of an arctic plant [64], but lower than that of Alcaligenes faecalis BCCM IC 2374 (Siderophore: 347 $\mu g\ ml^{-1}$) [65], suggesting that the plant growth promoting activity from different environments is totally different. The abundance of isolates producing more than 20 $\mu g\ ml^{-1}$ IAA suggested that the plant growth promoting ability of the isolates might assist in phytoremediating the soil.

The bacterial and fungal siderophores facilitate iron uptake in soil [10]. Iron chelated by siderophores is unavailable to plant pathogens resulting in an increase in plant health [13]. Metal–resistant siderophore-producing bacteria play important role in the successful survival and growth of plants in contaminated soil by alleviating metal toxicity and supplying nutrients for plant, and bacterial siderophore can bind metals other than iron [66], which may be the reason why microorganism can survive in the mine tailing soil contaminated by multi-metals. Bacterial siderophore should be beneficial to regulate availability of the abundant iron in the V-Ti magnetite mine tailing soil containing high concentration of iron. As bioaugmentation-assisted phytoextraction technology, the indigenous siderophore-producing bacteria can increase the phytoextraction rate that usually limits the use of phytoremediation methods [67]. Aside from their involvement in iron acquisition, siderophores have physiological roles of protecting some bacteria against the toxic effect of pyochelin by reducing reactive oxygen species [68], so the presence of siderophore-producing bacteria in the mine tailing can directly or indirectly promote bioremediation for the contaminated soil. Many isolates showed both IAA and siderophore production, implying that the characteristics of the indigenous bacteria are helpful in bioremediating the desert mine tailing area. Phytoremediation of metals was facilitated by PGPR by promoting plant growth and increasing the amount of metal taken up by plant [21],[69].

The heavy metal-resistance and plant growth-promoting activity are key characteristics for bacteria that are to be applied in metal phytoremediation. Therefore, analyzing these characteristics in an indigenous bacterial in contaminated sites is essential to provide significant information for developing effective bioremediation measures. Moreover, because both the structural and functional bacterial diversity are important indicators of soil health, evaluation for diversity of heavy metal-resistant bacteria and PGPR should be considered as the primary work for bioremediating soil contaminated by heavy metals. We showed that V-Ti magnetite mine tailing soil in Zhujiabaobao contained abundant bacteria that tolerate multiple heavy metals and have plant growth-promoting abilities. The results suggested that the indigenous bacteria in the soil have characteristics beneficial for remediating the contaminated mine tailing soil. To further study the phytoremediation approach, the plant growth promoting activity will be studied both in greenhouses and in situ in Zhujiabaobao.

AUTHOR CONTRIBUTIONS

Conceived and designed the experiments: XY QC. Performed the experiments: XY YL CZ HL JL WZ XK XL KZ QX. Analyzed the data: XY GY XZ QC. Contributed reagents/materials/analysis tools: XY QC. Contributed to the writing of the manuscript: XY.

REFERENCES

1. Neuberger JS, Mulhall M, Pomatto MC, Sheverbush J, Hassanein RS (1990) Health problems in Galena, Kansas: a heavy metal mining Superfund site. Sci Total Environ 94: 261–272. doi: 10.1016/0048-9697(90)90175-t.

2. Saunders J, Jastrzembski B, Buckey J, Enriquez D, MacKenzie T, et al. (2012) Hearing loss and heavy metal toxicity in a Nicaraguan mining community: Audiological results and case reports. Audiol Neurootol 18: 101–113. doi: 10.1159/000345470.

3.   Ohlander J, Huber SM, Schomaker M, Heumann C, Schierl R, et al. (2013) Risk Factors for Mercury Exposure of Children in a Rural Mining Town in Northern Chile. PloS one 8: e79756. doi: 10.1371/journal.pone.0079756.

4.   Boularbah A, Schwartz C, Bitton G, Morel JL (2006) Heavy metal contamination from mining sites in South Morocco: 1. Use of a biotest to assess metal toxicity of tailings and soils. Chemosphere 63: 802–810. doi: 10.1016/j.chemosphere.2005.07.079.

5.   Zhuang P, Lu H, Li Z, Zou B, McBride MB (2014) Multiple exposure and effects assessment of heavy metals in the population near mining area in South China. PloS one 9: e94484. doi: 10.1371/journal.pone.0094484.

6.   Mazej Z, Al Sayegh-Petkovšek S, Pokorny B (2010) Heavy metal concentrations in food chain of Lake Velenjsko jezero, Slovenia: an artificial lake from mining. Arch Environ Contam Toxicol 58: 998–1007. doi: 10.1007/s00244-009-9417-5.

7.   Valverde A, González-Tirante M, Medina-Sierra M, Santa-Regina I, García-Sánchez A, et al. (2011) Diversity and community structure of culturable arsenic-resistant bacteria across a soil arsenic gradient at an abandoned tungsten–tin mining area. Chemosphere 85: 129–134. doi: 10.1016/j.chemosphere.2011.06.025.

8.   Nielsen MN, Winding A (2002) Microorganisms as indicators of soil health. National Environmental Research Institute Denmark.

9.   Gupta AK, Verma SK, Khan K, Verma RK (2013) Phytoremediation using aromatic plants: a sustainable approach for remediation of heavy metals polluted sites. Environ Sci Technol 47: 10115–10116. doi: 10.1021/es403469c.

10.   Weyens N, van der Lelie D, Taghavi S, Newman L, Vangronsveld J (2009) Exploiting plant–microbe partnerships to improve biomass production and remediation. Trends Biotechnol 27: 591–598. doi: 10.1016/j.tibtech.2009.07.006.

11.   Fuentes-Ramirez LE, Jimenez-Salgado T, Abarca-Ocampo I, Caballero-Mellado J (1993) Acetobacter diazotrophicus, an indoleacetic acid producing bacterium isolated from sugarcane cultivars of Mexico. Plant Soil 154: 145–150. doi: 10.1007/bf00012519.

12. Jaroszuk-Ściseł J, Kurek E, Trytek M (2014) Efficiency of indoleacetic acid, gibberellic acid and ethylene synthesized in vitro by Fusarium culmorum strains with different effects on cereal growth. Biologia 69: 281–292. doi: 10.2478/s11756-013-0328-6.

13. Khan MS, Zaidi A, Wani PA, Oves M (2009) Role of plant growth promoting rhizobacteria in the remediation of metal contaminated soils. Environ Chem Lett 7: 1–19. doi: 10.1007/s10311-008-0155-0.

14. Tak HI, Ahmad F, Babalola OO (2013) Advances in the application of plant growth-promoting rhizobacteria in phytoremediation of heavy metals. Reviews of Rev Environ Contam Toxicol 223: 33–52. doi: 10.1007/978-1-4614-5577-6_2.

15. Leinhos V, Bergmann H (1995) Influence of auxin producing rhizobacteria on root morphology and nutrient accumulation of crops, pt. 2: root growth promotion and nutrient accumulation of maize (Zea mays L.) by inoculation with indole-3-acetic acid (IAA) producing pseudomonas strains and by exogenously applied IAA under different water supply conditions. Angewandte Botanik (Germany).

16. Lippmann B, Leinhos V, Bergmann H (1995) Influence of auxin producing rhizobacteria on root morphology and nutrient accumulation of crops, pt. 1: changes in root morphology and nutrient accumulation in maize (Zea mays L.) caused by inoculation with indole-3-acetic acid (IAA) producing Pseudomonas and Acinetobacter strains or IAA applied exogenously. Angewandte Botanik (Germany).

17. Rajkumar M, Freitas H (2008) Influence of metal resistant-plant growth-promoting bacteria on the growth of Ricinus communis in soil contaminated with heavy metals. Chemosphere 71: 834–842. doi: 10.1016/j.chemosphere.2007.11.038.

18. Olson JW, Mehta NS, Maier RJ (2001) Requirement of nickel metabolism proteinsHypA and HypB for full activity of both hydrogenase and urease in Helicobacter pylori. Mol Microbiol 39: 176–182. doi: 10.1046/j.1365-2958.2001.02244.x.

19. .Sakamoto T, Bryant DA (2001) Requirement of nickel as an essential micronutrient for the utilization of urea in the marine

cyanobacterium Synechococcus sp. PCC 7002. Plant Cell Physiol 42: 186. doi: 10.1264/jsme2.2001.177.

20. Mishra VK, Upadhyaya AR, Pandey SK, Tripathi B (2008) Heavy metal pollution induced due to coal mining effluent on surrounding aquatic ecosystem and its management through naturally occurring aquatic macrophytes. Bioresour Technol 99: 930–936. doi: 10.1016/j.biortech.2007.03.010.

21. Ma Y, Prasad M, Rajkumar M, Freitas H (2011) Plant growth promoting rhizobacteria and endophytes accelerate phytoremediation of metalliferous soils. Biotechnol Adv 29: 248–258. doi: 10.1016/j.biotechadv.2010.12.001.

22. Zhou M-F, Robinson PT, Lesher CM, Keays RR, Zhang C-J, et al. (2005) Geochemistry, petrogenesis and metallogenesis of the Panzhihua gabbroic layered intrusion and associated Fe–Ti–V oxide deposits, Sichuan Province, SW China. J Petrol 46: 2253–2280. doi: 10.1093/petrology/egi054.

23. Yanguo T, Shijun N, Xianguo T, Chengjiang Z, Yuxiao M (2002) Geochemical baseline and trace metal pollution of soil in Panzhihua mining area. Chinese J Geochemistry 21: 274–281. doi: 10.1007/bf02831093.

24. Yanguo T, Xianguo T, Shijun N, Chengjiang Z, Zhengqi X (2003) Environmental geochemistry of heavy metal contaminants in soil and stream sediment in Panzhihua mining and smelting area, Southwestern China. Chinese J Geochemistry 22: 253–262. doi: 10.1007/bf02842869.

25. Larsson MA, Baken S, Gustafsson JP, Hadialhejazi G, Smolders E (2013) Vanadium bioavailability and toxicity to soil microorganisms and plants. Environ Toxicol Chem 32: 2266–2273. doi: 10.1002/etc.2322.

26. Kužel S, Hruby M, Cígler P, Tlustoš P, Van Nguyen P (2003) Mechanism of physiological effects of titanium leaf sprays on plants grown on soil. Biol Trace Elem Res 91: 179–189. doi: 10.1385/bter:91:2:179.

27. Eckardt NA (2012) Pumping iron: conserved iron deficiency responses in the plant lineage. Plant Cell 24(10): 3855. doi: 10.1105/tpc.112.241010.

28. Venail PA, Vives MJ (2013) Positive effects of bacterial diversity on ecosystem functioning driven by complementarity effects in

a bioremediation context. PloS one 8: e72561. doi: 10.1371/journal.pone.0072561

29. Li Z, Zhang G, Liu Y, Wan K, Zhang R, et al. (2013) Soil nutrient assessment for urban ecosystems in Hubei, China. PloS one 8: e75856. doi: 10.1371/journal.pone.0075856.

30. Zhang S, Chen M, Li T, Xu X, Deng L (2010) A newly found cadmium accumulator–Malva sinensis Cavan. J Hazard Mater 173: 705–709. doi: 10.1016/j.jhazmat.2009.08.142.

31. Chang S, Hsu H, Cheng J, Tseng C-P (2011) An efficient strategy for broad-range detection of low abundance bacteria without DNA decontamination of PCR reagents. PloS one 6: e20303. doi: 10.1371/journal.pone.0020303.

32. Martin B, Humbert O, Camara M, Guenzi E, Walker J, et al. (1992) A highly conserved repeated DNA element located in the chromosome of Streptococcus pneumoniae. Nucleic Acids Res 20: 3479–3483. doi: 10.1093/nar/20.13.3479.

33. Tacão M, Alves A, Saavedra MJ, Correia A (2005) BOX-PCR is an adequate tool for typing Aeromonas spp. Antonie van Leeuwenhoek 88: 173–179. doi: 10.1007/s10482-005-3450-9.

34. Miller CS, Handley KM, Wrighton KC, Frischkorn KR, Thomas BC, et al. (2013) Short-read assembly of full-length 16S amplicons reveals bacterial diversity in subsurface sediments. PloS one 8: e56018. doi: 10.1371/journal.pone.0056018.

35. Wilson KH, Blitchington R, Greene R (1990) Amplification of bacterial 16S ribosomal DNA with polymerase chain reaction. J Clin Microbiol 28: 1942–1946.

36. Benson DA, Cavanaugh M, Clark K, Karsch-Mizrachi I, Lipman DJ, et al.. (2012) GenBank. Nucleic acids research: gks1195.

37. **37.**Tamura K, Stecher G, Peterson D, Filipski A, Kumar S (2013) MEGA6: molecular evolutionary genetics analysis version 6.0. Mol Biol Evol 30: 2725–2729. doi: 10.1093/molbev/mst197.

38. Bric JM, Bostock RM, Silverstone SE (1991) Rapid in situ assay for indoleacetic acid production by bacteria immobilized on a nitrocellulose membrane. Appl Environ Microb 57: 535–538.

39. Patten CL, Glick BR (2002) Role of Pseudomonas putida indoleacetic acid in development of the host plant root system. Appl Environ Microb 68: 3795–3801. doi: 10.1128/aem.68.8.3795-3801.2002.

40. Schwyn B, Neilands J (1987) Universal chemical assay for the detection and determination of siderophores. Analytical biochemistry 160: 47–56. doi: 10.1016/0003-2697(87)90612-9.

41. Murugappan R, Rekha S, Thirumurugan R (2006) Characterization and quantification of siderophores produced by Aeromonas hydrophila isolated from Cyprinus carpio. Pak J Biol Sci 9: 437–440. doi: 10.3923/pjbs.2006.437.440.

42. Rodríguez L, Ruiz E, Alonso-Azcárate J, Rincón J (2009) Heavy metal distribution and chemical speciation in tailings and soils around a Pb–Zn mine in Spain. J Environ Manage 90: 1106–1116. doi: 10.1016/j.jenvman.2008.04.007.

43. Tripathi BM, Kim M, Singh D, Lee-Cruz L, Lai-Hoe A, et al. (2012) Tropical soil bacterial communities in Malaysia: pH dominates in the equatorial tropics too. Microbial Ecol 64: 474–484. doi: 10.1007/s00248-012-0028-8.

44. Lauber CL, Hamady M, Knight R, Fierer N (2009) Pyrosequencing-based assessment of soil pH as a predictor of soil bacterial community structure at the continental scale. Appl Environ Microb 75: 5111–5120. doi: 10.1128/aem.00335-09.

45. Brady NC, Weil RR (1996) The nature and properties of soils. Prentice-Hall Inc.

46. Shacklette HT, Boerngen JG (1984) Element concentrations in soils and other surficial materials of the conterminous United States.

47. Abreu CAd, van Raij B, Abreu MFd González AP (2005) Routine soil testing to monitor heavy metals boron. Sci Agr 62: 564–571. doi: 10.1590/s0103-90162005000600009.

48. Peralta-Videa J, Gardea-Torresdey J, Gomez E, Tiemann K, Parsons J, et al. (2002) Effect of mixed cadmium, copper, nickel and zinc at different pHs upon alfalfa growth and heavy metal uptake. Environ Pollut 119: 291–301. doi: 10.1016/s0269-7491(02)00105-7.

49. Das S, Mishra J, Das SK, Pandey S, Rao DS, et al. (2014) Investigation on mechanism of Cr (VI) reduction and removal by Bacillus amyloliquefaciens, a novel chromate tolerant bacterium isolated from chromite mine soil. Chemosphere 96: 112–121. doi: 10.1016/j.chemosphere.2013.08.080.

50. Govarthanan M, Lee K-J, Cho M, Kim JS, Kamala-Kannan S, et al. (2013) Significance of autochthonous Bacillus sp. KK1 on biomineralization of lead in mine tailings. Chemosphere 90: 2267–2272. doi: 10.1016/j.chemosphere.2012.10.038.

51. Cheng Y, Yan F, Huang F, Chu W, Pan D, et al. (2010) Bioremediation of Cr (VI) and immobilization as Cr (III) by Ochrobactrum anthropi. Environ Sci Technol 44: 6357–6363. doi: 10.1021/es100198v.

52. Pandey S, Ghosh PK, Ghosh S, De TK, Maiti TK (2013) Role of heavy metal resistantOchrobactrum sp. and Bacillus spp. strains in bioremediation of a rice cultivar and their PGPR like activities. J Microbiol 51: 11–17. doi: 10.1007/s12275-013-2330-7.

53. Wasi S, Tabrez S, Ahmad M (2013) Use of Pseudomonas spp. for the bioremediation of environmental pollutants: a review. Environ Monit Assess 185: 8147–8155. doi: 10.1007/s10661-013-3163-x.

54. Ike A, Sriprang R, Ono H, Murooka Y, Yamashita M (2007) Bioremediation of cadmium contaminated soil using symbiosis between leguminous plant and recombinant rhizobia with the MTL4 and the PCS genes. Chemosphere 66: 1670–1676. doi: 10.1016/j.chemosphere.2006.07.058.

55. Sriprang R, Hayashi M, Yamashita M, Ono H, Saeki K, et al. (2002) A novel bioremediation system for heavy metals using the symbiosis between leguminous plant and genetically engineered rhizobia. J Biotechnol 99: 279–293. doi: 10.1016/s0168-1656(02)00219-5.

56. Pumpel T, Paknikar KM (2001) Bioremediation technologies for metal-containing wastewaters using metabolically active microorganisms. Adv Appl microbiol 48: 135–171. doi: 10.1016/s0065-2164(01)48002-6.

57. Xie X, Fu J, Wang H, Liu J (2010) Heavy metal resistance by two bacteria strains isolated from a copper mine tailing in China. Afr J Biotechnol 9: 4056–4066.

58. Bafana A, Krishnamurthi K, Patil M, Chakrabarti T (2010) Heavy metal resistance inArthrobacter ramosus strain G2 isolated from mercuric salt-contaminated soil. J Hazard Mater 177: 481–486. doi: 10.1016/j.jhazmat.2009.12.058.

59. Joseph B, Ranjan Patra R, Lawrence R (2012) Characterization of plant growth promoting rhizobacteria associated with chickpea (Cicer arietinum L.). Intl J Plant Prod 1: 141–152.

60. Roszak D, Colwell R (1987) Survival strategies of bacteria in the natural environment. Microbiol Rev 51: 365.

61. Ahmad F, Ahmad I, Khan M (2008) Screening of free-living rhizospheric bacteria for their multiple plant growth promoting activities. Microbiol Res 163: 173–181. doi: 10.1016/j.micres.2006.04.001.

62. Leinhos V, Bergmann H (1995) Influence of auxin producing rhizobacteria on root morphology and nutrient accumulation of maize (Zea mays L.) by inoculation with indol-3-acetic acid (IAA) producing Pseudomonas strains and by exogenously applied IAA under different water supply conditions. Angew Bot 69: 37–42.

63. Kuffner M, De Maria S, Puschenreiter M, Fallmann K, Wieshammer G, et al. (2010) Culturable bacteria from Zn-and Cd-accumulating Salix caprea with differential effects on plant growth and heavy metal availability. J App Microbiol 108: 1471–1484. doi: 10.1111/j.1365-2672.2010.04670.x.

64. Xie H, Pasternak J, Glick BR (1996) Isolation and characterization of mutants of the plant growth-promoting rhizobacterium Pseudomonas putida GR12–2 that overproduce indoleacetic acid. Curr Microbiol 32: 67–71. doi: 10.1007/s002849900012.

65. Sayyed R, Chincholkar S (2010) Growth and siderophores production in Alcaligenes faecalis is regulated by metal ions. Indian H Microbiol 50: 179–182. doi: 10.1007/s12088-010-0021-1.

66. Rajkumar M, Ae N, Prasad MN, Freitas H (2010) Potential of siderophore-producing bacteria for improving heavy metal phytoextraction. Trends Biotechnol 28: 142–149. doi: 10.1016/j.tibtech.2009.12.002.

67. Braud A, Jezequel K, Bazot S, Lebeau T (2009) Enhanced phytoextraction of an agricultural Cr- and Pb-contaminated soil by bioaugmentation with siderophore-producing bacteria. Chemosphere 74: 280–286. doi: 10.1016/j.chemosphere.2008.09.013.

68. Adler C, Corbalán NS, Seyedsayamdost MR, Pomares MF, de

Cristóbal RE, et al. (2012) Catecholate siderophores protect bacteria from pyochelin toxicity. PloS one 7: e46754. doi: 10.1371/journal.pone.0046754.

69. Glick BR (2010) Using soil bacteria to facilitate phytoremediation. Biotechnol Adv 28: 367–374. doi: 10.1016/j.biotechadv.2010.02.001.

Citations

CHAPTER 1

Raymond A. Wuana and Felix E. Okieimen, "Heavy Metals in Contaminated Soils: A Review of Sources, Chemistry, Risks and Best Available Strategies for Remediation," ISRN Ecology, vol. 2011, Article ID 402647, 20 pages, 2011. doi:10.5402/2011/402647.

CHAPTER 2

Jarrod O. Miller, Anastasios D. Karathanasis, and Christopher J. Matocha, "In Situ Generated Colloid Transport of Cu and Zn in Reclaimed Mine Soil Profiles Associated with Biosolids Application," Applied and Environmental Soil Science, vol. 2011, Article ID 762173, 9 pages, 2011 doi:10.1155/2011/762173.

CHAPTER 3

Cleide S. T. Araújo, Dayene C. Carvalho, Helen C. Rezende, Ione L. S. Almeida, Luciana M. Coelho, Nívia M. M. Coelho, Thiago L. Marques and Vanessa N. Alves (2013). Bioremediation of Waters Contaminated with Heavy Metals Using Moringa oleifera Seeds as Biosorbent, Applied Bioremediation - Active and Passive Approaches, Dr. Yogesh Patil (Ed.), ISBN: 978-953-51-1200-6, InTech, DOI: 10.5772/56157.

CHAPTER 4

Nazan Kuter (2013). Reclamation of Degraded Landscapes due to Opencast Mining, Advances in Landscape Architecture, Dr. Murat Ozyavuz (Ed.), ISBN: 978-953-51-1167-2, InTech, DOI: 10.5772/55796.

CHAPTER 5

Santosh Kumar Sarkar, Paulo J.C. Favas, Dibyendu Rakshit and K.K. Satpathy (2014) Geochemical Speciation and Risk Assessment of Heavy Metals in Soils and Sediments, Environmental Risk Assessment of Soil Contamination, Dr. Maria C. Hernandez Soriano (Ed.), ISBN: 978-953-51-1235-8, InTech, DOI: 10.5772/57295.

CHAPTER 6

Kpan, J. , Opoku, B. and Gloria, A. (2014) Heavy Metal Pollution in Soil and Water in Some Selected Towns in Dunkwa-on-Offin District in the Central Region of Ghana as a Result of Small Scale Gold Mining. Journal of Agricultural Chemistry and Environment, **3**, 40-47. doi: 10.4236/jacen.2014.32006.

CHAPTER 7

Yu X, Li Y, Zhang C, Liu H, Liu J, et al. (2014) Culturable Heavy Metal-Resistant and Plant Growth Promoting Bacteria in V-Ti Magnetite Mine Tailing Soil from Panzhihua, China. PLoS ONE 9(9): e106618. doi:10.1371/journal.pone.0106618.

Index